From Shockwaves to Cavitation: The Wild Ride of Multi-Phase Flows

Naviya

Contents

1 Introduction **1**

  1.1 Background 1

  1.2 Objectives 4

  1.3 Original Contributions in the work 5

  1.4 Plan of Development 5

2 Basic Mathematical Formulation **6**

  2.1 Introduction 6

  2.2 Assumptions 6

    2.2.1 Continuum Hypothesis 6

    2.2.2 Fluid Physics considered 7

  2.3 Control Volume Analysis 7

  2.4 Multi-phase flow 9

2.4.1 Two-Phase Flow 9

2.4.2 The Stiffened Gas EOS 11

2.4.3 Interface properties (mixed cells) 13

2.5 Summary of governing equations 15

3 Numerical Methodology   **16**

3.1 Introduction 16

3.2 Discretisation of spatial and temporal terms 17

3.2.1 FV Median Dual Cell mesh Variant 17

3.2.2 Second-order spatial Reconstruction 19

3.2.3 Temporal Integration and stability 21

3.2.4 Eigenvalues and vectors 22

3.3 Multi-dimensional two-phase HLLC solver 24

3.3.1 Consistency conditions 25

3.3.2 Inviscid HLLC-CICSAM edge flux 26

3.3.3 Surface tension source terms 31

3.3.4 Semi-Analytical Proof of Well-Balanced Scheme 33

3.4 Curvature Reconstruction 34

3.5 VoF equation 35

3.6 Summary of Algorithm 38

4 Numerical Test-cases   **40**

4.1 Introduction 40

4.2 1-D Test-Cases 40

4.2.1 Sod-Shock tube 41

4.2.2 Interface only 42

4.2.3 Gas-Liquid Shock-tube 43

4.3 2-D test-cases 45

4.3.1 Advecting bubble in an oblique velocity field 45

4.3.2 Under-water Explosion 49

4.3.3 Shock-Bubble Interaction 52

4.3.4 Spurious currents in a static test-case 54

4.3.5 Oscillating bubble 56

4.3.6 Rayleigh-Plesset collapse problem 58

5   Conclusion   **63**

   5.1   Summary   63

   5.2   Recommendations and future work   64

Appendices   **66**

A   **67**

   A.1   Derivation of thermal EOS   67

   A.2   Derivation of HLLC intermediate star-state   69

   A.3   Analytical solution for 1-D test cases   71

   A.4   Derivation of 2-D Rayleigh-Plesset Equation   71

   A.5   Runge-Kutta fourth-order - Rayleigh-Plesset model   72

   A.6   Ethics Clearance   74

Chapter 1

Introduction

1.1 Background

High-speed multi-phase compressible flow induced by blast or shock waves is of interest
to both basic science and engineering. For example, when a sample of a solid metal is
subjected to a high-power laser beam, the large negative pressures created in the metal
lead to its instant melting followed by micro-spalling [6,7]. In an under-water explosion,
the detonation of an unconfined charge leads to the growth of a stable gas bubble [8]. In
these examples, the admittance of compressible effects in the liquid medium in addition
to the liquid-gas interface motion are key fluid physics phenomena.

In recent years, the importance of accurately modelling these effects has been further
highlighted in such work involving shock-wave induced liquid fragmentation [8–10]. For
instance, Milne et al. [9] considered the interaction of a highly charged explosive with a
spherical liquid geometry. Several key physics processes are described in their work that
demonstrate the significance of accurately capturing the sonic waves and the liquid-gas
interface. First, an outgoing shock wave compresses the water molecules into a dense
region. Subsequently, this shock wave propagates through the water layer where it is
driven into air at the free surface. As a consequence, a release wave is pushed back
into the water medium. This rarefaction wave is a tensile wave and leaves a region of
cavitated water behind. As these shock waves traverse the water layer, they lead to
the compression and expansion of the liquid phase resulting in the violent growth of a
heterogeneous mixture (gas and liquid). Intrinsically, the sonic characteristics of the flow
are inherently linked to the dilatation of the liquid phase in the liquid-gas mixture.

In addition to liquid compressibility, surface tension plays a defining role in terms of bubble growth and liquid fragmentation. Such a conclusion can be drawn from the work by Frost [11]. Notably, in the cited work, the dynamic response of high-speed liquid-gas flows is dependent on several effects, viz. the nature of the fluid itself, shock-interface interaction, vaporisation, surface tension effects, transient forces and heat or mass transfer between liquid and gas phases. Three mechanisms are proposed in their work to explain the behaviour of shock-induced liquid fragmented flows. The first relates to a shock driven multi-phase instability, which is analogous to Rayleigh Taylor or Richtmyer-Meshkov instability, where a denser fluid (water) is slowed down by a lighter fluid (air). The second is characterised by the acceleration of the liquid phase, where the latter is compacted into a dense region and deforms inelastically leading to the formation of jet structures as discussed by Milne et al. [9]. The third is similar to the second mechanism but results in the formation of conically shaped jets. In particular, for such flows involving instabilities, as demonstrated by Durand et al. [12], surface tension has a major impact on the size and velocity of the jetted particles. Hence, the qualitative understanding of shock-induced multi-phase fragmented flows relies heavily on accounting for both the compressible nature of the liquid phase and surface tension effects.

Over the last decade, extensive work has been done to model such interaction. Three methods are outlined in literature, viz. the Finite Difference (FD), Finite Element (FE) and Finite Volume (FV) method. For instance, Ghoshal et al. [13] modelled the interaction of an under-water explosion where non-linear compressibility effects of the liquid are accounted for using an orthogonal FD method with a Lagrangian formulation of Euler's equation for compressible flow. Shin et al. [14] used a Lagrangian-Eulerian coupled FE method to model the interaction of the under-water explosion with a surrounding spherical structure. Milne et al. [9] used both a continuum and a two-phase model via a FV method to simulate the interaction of a highly charged explosive with a thin layer of water. Yet, in most of these aforementioned articles, either compressibility or surface tension effects are neglected.

From the FV approach, foundational work has been laid to model multi-phase compressible flow. These may be broadly divided into three schools of thought viz., the two-phase [15, 16] (7 equations), reduced [17] (5 equations) or the homogeneous flow models [18]. Each model presents strengths and weaknesses which will be further critiqued in Chapter 2. Numerically, to obtain the so-called inter-cell Godunov flux [19, 20] used to solve the governing equations, a number of methods have been employed over the last decade. These can be broadly categorised into the following families of solvers, viz. the Riemann solvers [18, 21], the Godunov type schemes [22, 23] and the semi-implicit projection methods [24, 25]. For example, Shyue [18] employed a homogeneous gamma based model to capture the propagation of shock waves using the Riemann solver of Roe. Johnsen et al. [26] implemented an adaptation of the Harten's, Lax, Leer and Contact (HLLC) [21] approximate Riemann solver for modelling two-phase compressible flow. However, as it will be shown in this work, when applied to track the liquid-gas interface, Riemann methods result in smearing leading to the loss of the sharp characteristics of the inter-facial flow. This has significant implications on the ability to compute the interface curvature accurately for the purpose of modelling surface tension effects.

In this regard, different approaches [1, 22, 24, 27–31] have been proposed in literature to account for surface tension for compressible flow. For instance, Perigaud et al. [22] employed a Godunov type scheme via a Diffuse Interface Method for modelling capillary effects. Though their approach exploit the conservative properties of the Euler system via the inclusion of a surface tension energy term, again, they suffer from the smearing of the liquid-gas interface [31].

Further, a number of interface handling methods are outlined in literature to circumvent the smearing. One class of such methods involves tracking the interface explicitly either by using a Lagrangian or an Arbitrary Lagrangian Eulerian (ALE) reference frame. For example, Daude et al. [32] extended the HLLC solver to an ALE formulation in an attempt to keep the interface sharp. However, in their work, no surface tension effects were accounted for. Moreover, typical Lagrangian or ALE methods are known to be limited by mesh distortions [33] and therefore incur additional complexities which is computationally expensive.

Other interface handling methods include the use of interface-sharpening techniques. For example, in the work by Garrick et al. [1], a compression interface scheme was proposed to allow for the inclusion of surface tension effects. He et al. [34] adopted a similar approach where a pressure intermediate equilibrium state similar to a Riemann flux was derived but the latter did not include any surface tension effects. Moreover, such interface sharpening methods [1, 34] are not based on a liquid availability criteria [35]. This will be further expanded on in Chapter 3. Other notable work for surface tension in compressible flow have involved the use of the Ghost-Fluid-Method (GFM) [36]. Such an approach for interface tracking was adopted by Fechter et al. [29]. In their work, the original Hartens, Lax and Leer (HLL) solver was extended to account for both phase transition and surface tension effects. Nevertheless, this method typically requires extra physics to maintain the robustness of the solver [33].

The final class of interface handling methods is that of the sharp interface techniques [37]. This class can broadly divided into two families, viz. the interface tracking and interface capturing method. Interface tracking includes the Marker And Cell (MAC) particles [38] and the Front tracking method [39]. The interface capturing methods are the Level Set method [40, 41] and the Volume-of-Fluid (VoF) [42] method. The interface tracking methods and the Level Set method are not strictly volume conservative [43, 44]. In contrast, VoF advective schemes have been shown [35, 45–47] to maintain the sharpness of the interface in a bounded and volume conservative manner.

Only recently has the use of such VoF methods gained pace in the field of compressible flow modelling [24, 30, 31, 48]. Fuster et al. [24] altered the unified compressible-incompressible semi-implicit projection method proposed by Xiao [49] and extended the formulation to allow for the inclusion of surface tension effects. Notably, in the latter work, an accurate geometric VoF method [45] was implemented to track the interface. The interface accuracy is due to geometrically accounting for liquid availability in mixed cells when computing phase fluxes. Yet, historically, all VoF methods which account for liquid availability have mostly been implemented in conjunction with pressure-based fractional step methods [24, 43, 50–52]. Notably, these have widely been pioneered in the context of incompressible or weakly compressible flow [43, 50–56].

To the author's knowledge, only [30, 31, 48] for compressible flow successfully combine a VoF method with a typical Riemann or Godunov type solver. Shyue [48] implemented a geometric VoF method with the Riemann Solver of Roe but, however, reconstructed conserved variables and neglected surface tension effects. Corot et al. [31] employed a similar geometric VoF approach to account for surface tension. However, their work required the use of a newly derived Godunov type scheme and not an existing Riemann solver. Jibben et al. [30] employed the VoF piece-wise linear interface calculation to track the interface with the HLLC solver on an adaptive refined mesh in the presence of surface tension effects. However, their proposed VoF scheme is restricted to Cartesian grids.

Hence, there is a need to further expand and develop the framework to couple an existing Riemann type flow solver with any VoF scheme. This is in a manner that allows for the efficient capture of the shock and rarefaction waves while maintaining the geometric integrity of the interface for the purpose of curvature computation for surface tension modelling. This leads to the objectives of this work.

1.2 Objectives

An existing computer code ELEMENTAL® [50, 53, 55, 57] was employed as a platform for this book, which is limited to weakly compressible VoF via a split pressure-projection solver. Hence, the objectives of this work are,

- to develop a homogeneous two-phase compressible flow model,

- to combine the above compressible flow model with an algebraic VoF method adaptable to un-structured grids,

- to include surface tension effects in the developed model,

- to demonstrate the validity of the proposed solver on existing numerical test-cases in 1-D and 2-D.

Finally, this project hopes to lay down ground work for modelling the dynamic fragmentation of liquid media.

1.3 Original Contributions in the work

This work presents a novel modified HLLC multi-phase solver that allows for the inclusion of surface tension effects similar to [1]. However, in contrast to the latter cited article, the VoF method is based on the availability criteria while surface tension is accounted for accurately on structured grids via the celebrated height function method [3, 4] and the convolution method on unstructured grids. CICSAM [58] is employed for the VoF equation. This is the first time that such a combination has been successfully implemented.

1.4 Plan of Development

This work is organised as follows. In Chapter 2, the basic mathematical formulation for the governing equations employed for modelling the two-phase inviscid compressible flow are presented. These conservation equations are first written in weak integral form and re-formulated for homogeneous compressible flow. In Chapter 3, the numerical method implemented to solve the governing equations is described. In Chapter 4, the numerical test-cases to validate the proposed model are presented. Finally, in Chapter 5, the conclusions and recommendations of this work are outlined.

Chapter 2

Basic Mathematical Formulation

2.1   Introduction

In this chapter, the governing equations for describing the dynamics of an immiscible two-phase compressible flow a re p resented. T he m odel i s b ased o n e xisting numerical work [16, 18, 20]. A control volume analysis is used to obtain the basic mathematical formulation for the conservation of physical quantities such as mass, momentum and energy. Detailed derivations of these conservation equations can be found in the following Fluid Mechanics textbooks [20, 59, 60] and are not included in this book.

Here, the governing equations are written in weak integral form to allow for the treatment of discontinuities in the flow. The closure thermodynamic conditions are set via the Stiffened Gas Equation Of State [61] (EOS). In particular, both caloric and thermal relationships are established for the mixture based on two parameters, viz. pressure and density for the former and pressure and temperature for the latter. In the next section, the underlying assumptions used to derive the model are detailed.

2.2   Assumptions

2.2.1   Continuum Hypothesis

Any fluid is made up of a large number of molecules that are constantly interacting with each other. Indeed, matter is an aggregation of molecules within a certain volume. In a gas, the molecules are far apart while in a liquid, the molecules are closer to each other. The randomised collision and arrangement of individual molecules within this volume give rise to common physical quantities such as pressure or fluid density.

Treating this fluid as a continuous distribution of molecules rather than analysing the latter on a molecular level is known as the continuum hypothesis. Malan [52] points out to two key metrics to determine when the continuum hypothesis fails based on literature by White [59] and Kundu et al. [60]. The first relates to the volumetric scale on which the molecules are acting. In particular, there exists a certain limiting volume δV^* after which individual molecular variations become significant [59] i.e. the volume is too small to ignore intermolecular interaction. White [59] states that this lower limit is approximately 10^{-9} mm^3 for both gases and liquids at atmospheric pressure. Similarly, Tryggvason et al. [62] state that for dilute gases, the mean distance between such intermolecular collisions known as the mean free path of molecules is the limiting scale. As per Kundu et al. [60], the mean free path for standard atmospheric air is approximately 5×10^{-8} m. For the test-cases considered in this work, the volume is well within the aforementioned limits such that the flow can be described by its continuous distribution of fluid molecules.

2.2.2 Fluid Physics considered

For multi-phase flow, the free surface will be treated as sharp. In particular, the two fluids will be considered to be immiscible i.e. there is no dilution of mass from the liquid to the gas or vice versa. In other words, since the continuum hypothesis is applicable, the interface is assumed to have vanishing thickness [62]. Further, to limit the scope of the project, phase change is not accounted for. Moreover, as per [62], long range forces between molecules such as electromagnetic forces between charged fluids are ignored. In particular, air is modelled as an ideal gas at normal temperature and pressure, while intermolecular forces such as Van Der Waal (VDW) forces in real gases and liquids are included as an averaged empirical pressure via the EOS. Capillary effects such as surface tension are accounted for as volumetric forces. Finally, viscous and gravitational effects are ignored in this work.

2.3 Control Volume Analysis

Consider a Control Volume (CV) $\mathcal{V}$ at time t moving in some flow field $\boldsymbol{u}$ bounded by a Control Surface (CS) $\mathcal{S}$ as shown in Figure 2.1.

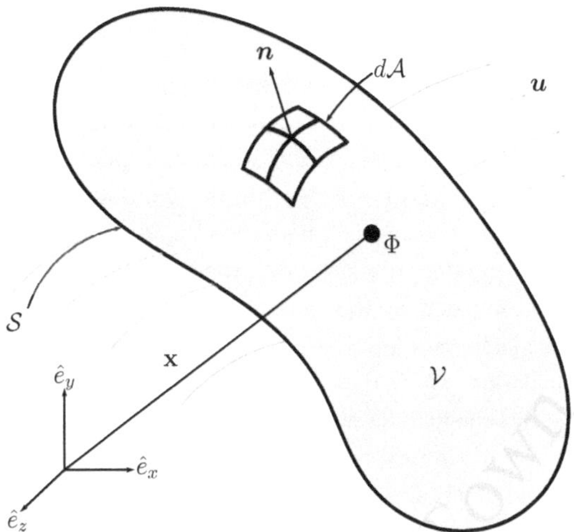

Figure 2.1: Schematic representation of Control Volume.

By applying the Reynolds' Transport Theorem, for a conserved scalar property Φ contained inside this CV, the following is true:

$$\frac{D}{Dt}\int_{\mathcal{V}}\rho\Phi d\mathcal{V} = \int_{\mathcal{V}}\frac{\partial(\rho\Phi)}{\partial t}d\mathcal{V} + \oint_{\mathcal{S}}\rho\Phi\boldsymbol{u}\cdot\boldsymbol{n}d\mathcal{A} = 0, \tag{2.1}$$

where, in this work, the material (Lagrangian) derivative is defined with respect to Eulerian derivatives as follows:

$$\frac{D}{Dt} = \frac{\partial}{\partial t} + \boldsymbol{u}\cdot\nabla,$$

with $\nabla = \frac{\partial}{\partial \mathrm{x}_j}\hat{\boldsymbol{e}}_j$ and where $\hat{\boldsymbol{e}}_j$ denotes a unit direction in a non-inertial Cartesian reference frame. Further, ρ is the density, $d\mathcal{A}$ is the area, $\boldsymbol{n}$ is the outward pointing unit normal of a segment on the CS, $\boldsymbol{u} = \boldsymbol{u}(\mathbf{x}, t)$ is the fluid velocity with the coordinate vector $\mathbf{x} = \mathrm{x}_j\hat{\boldsymbol{e}}_j$, and t is time in seconds. Finally, using the Divergence theorem, Equation (2.1) may be re-written in weak form as:

$$\frac{D}{Dt}\int_{\mathcal{V}}\rho\Phi d\mathcal{V} = \int_{\mathcal{V}}\frac{\partial(\rho\Phi)}{\partial t} + \nabla\cdot(\rho\Phi\boldsymbol{u})\,d\mathcal{V} = 0,$$

where the nomenclature has been previously defined. The above forms the basis for the conservation of all physical properties present in the liquid-gas mixture.

2.4   Multi-phase flow

Multi-phase compressible and incompressible flow models have been extensively studied over the past three decades. These may be broadly divided into three families of multi-phase modelling. First, the two-phase model where each phase is treated explicitly by its own set of equations. In this regard, the Baer and Nunziato type model [15] and the Abgrall and Saurel two-phase model [16] have been popular [63–66]. A second approach reduces the two energy equations to a single one, which is equally valid. Examples include the reduced five-equation model by Kapila et al. [17] used in [22, 32, 67].

The third, and final approach, involves the use of a homogeneous flow model known as the "one-fluid" formulation, where an equilibrium is assumed to exist between the liquid and the gas phases. This one-fluid formulation has been extensively used in incompressible flow [43, 50–52], and recently in compressible shock modelling [24, 48] of liquid-gas systems. The validity of such a model relies heavily on the time scale on which the flow reaches equilibrium [50].

Moreover, two distinct models are described in literature to track material properties in the two-phase flow. First, a gamma based model [18] where each thermodynamic quantity is passively transported in the flow. These are computationally expensive due to solving additional set of transport equations. A second method involves the use of the popular volume fraction model [1, 32, 35, 43, 48, 52, 63]. In this method, one phase is tracked and the properties are computed as a weighting of each fluid with respect to the average cell-volume. In this work, due to its simple and elegant implementation, a homogeneous VoF model is used for the inviscid modelling of the liquid-gas flow. The emphasis will be laid on the sharp capture of sonic shocks as well as the liquid-gas interface.

2.4.1   Two-Phase Flow

The employed governing equations for inviscid immiscible two-phase flow are based on that by Saurell and Abgrall [16]. In particular, a homogeneous flow model is derived from their work where the volume fraction for an arbitrary computational cell l (Ω_l), denoted by α_l, is first defined as the ratio of the cell average of the volume occupied by the tracked phase ($\mathcal{V}_i$) to the total volume of the cell $\mathcal{V}_l$:

$$\alpha_l = \frac{\mathcal{V}_i}{\mathcal{V}_l}.$$

The transport equation for the volume fraction field is obtained from its material derivative and can be written in semi-conservative form as:

$$\frac{\partial \alpha}{\partial t} + \nabla \cdot (\alpha \boldsymbol{u}) = \alpha \nabla \cdot \boldsymbol{u},$$

where $\boldsymbol{u}$ denotes a cell-average velocity. This formulation has been used by Weymouth et al. [45] and Fuster et al. [24] where the right-hand-side is referred to as the dilatation term (compression and expansion of the tracked phase).

Further, by defining a cell-average density ρ, pressure p, in the absence of gravitational, viscous effects and heat or mass transfer, the homogeneous governing equations then read in weak integral form:

$$\int_\Omega \frac{\partial}{\partial t} (\rho) \, d\mathcal{V} + \oint_{\partial \Omega} \rho \boldsymbol{u} \cdot \boldsymbol{n} d\mathcal{A} = 0,$$

$$\int_\Omega \frac{\partial}{\partial t} (\rho \boldsymbol{u}) \, d\mathcal{V} + \oint_{\partial \Omega} (\rho \boldsymbol{u} \otimes \boldsymbol{u} + p\boldsymbol{I}) \cdot \boldsymbol{n} d\mathcal{A} = \int_\Omega \boldsymbol{f}_\sigma d\mathcal{V},$$

$$\int_\Omega \frac{\partial}{\partial t} (\rho E) \, d\mathcal{V} + \oint_{\partial \Omega} (\rho E + p) \, \boldsymbol{u} \cdot \boldsymbol{n} d\mathcal{A} = \int_\Omega \boldsymbol{f}_\sigma \cdot \boldsymbol{u} d\mathcal{V},$$

where ρ is weighted as:

$$\rho = \alpha \rho_1 + (1 - \alpha) \rho_2,$$

with $i \in \{1, 2\}$ denoting the liquid and gas phase respectively. Moreover, $\boldsymbol{f}_\sigma$ are line forces representing surface tension or capillary effects, $\boldsymbol{I}$ is the identity matrix and E is the specific total energy. This total energy is expressed as the sum of the kinetic and internal energy ρe:

$$\rho E = \frac{1}{2} \rho \boldsymbol{u} \cdot \boldsymbol{u} + \rho e,$$

where ρe is computed using:

$$\rho e = \alpha \rho_1 e_1 + (1 - \alpha) \rho_2 e_2, \tag{2.2}$$

where $\rho_i e_i$ is obtained using the Stiffened Gas EOS [61]. Finally, with the introduction of the VoF method, the capillary line force $\boldsymbol{f}_\sigma$ is interpreted as a volumetric force as per the Continuum Surface Force (CSF) method [5] yielding:

$$\boldsymbol{f}_\sigma = \sigma \kappa \boldsymbol{n} \delta_s \approx \sigma \kappa \nabla \alpha,$$

with κ being the interface curvature, δ_s is the surface Dirac function which activates the surface tension term only at the interface and σ denotes the surface tension coefficient.

2.4.2 The Stiffened Gas EOS

The choice for the EOS used for the mixture model is a key component to retain the hyperbolic nature of the governing equations.

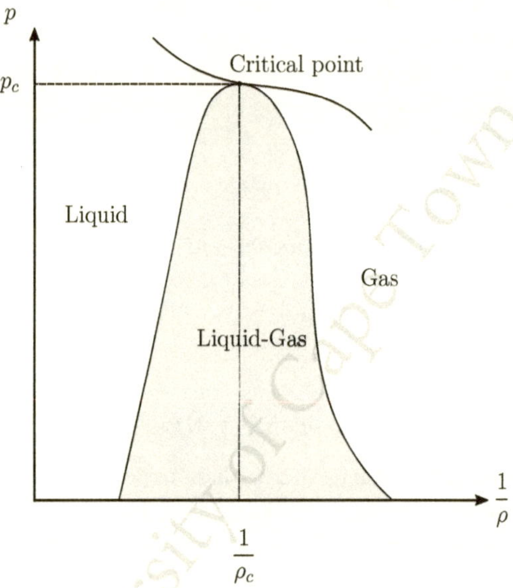

Figure 2.2: Schematic representation of saturation curves.

In particular, the loss of hyperbolicity may lead to an imaginary (un-physical) acoustic velocity [61] in the mixture zone (see Liquid-Gas region - Figure 2.2). In this work, the system of equations is closed by the caloric relationship $p = p\,(\rho, e)$ established via the Stiffened Gas EOS. This EOS has been widely used for multi-phase compressible flow modelling [1, 18, 24, 32, 48, 63] and is an approximation derived based on the saturation curves. In particular, this approximation is an extension to the ideal gas law where an empirical pressure p_∞ is included to model the intermolecular forces or VDW forces present in a real gas or liquid. The Stiffened Gas EOS for some phase i written in its caloric form reads:

$$\rho_i e_i = \frac{p + \gamma_i p_{\infty_i}}{\gamma_i - 1},\tag{2.3}$$

where γ_i is the adiabatic coefficient given as:

$$\gamma_i = \frac{c_{p_i}}{c_{v_i}}.$$

Here, c_{p_i} and c_{v_i} respectively denote the specific heat at constant pressure and temperature. This EOS is applicable for two-phase immiscible flow where the temperature and pressure are below the critical point. The thermal formulation of the EOS as per Le Métayer et al. [61] relating the internal energy to the pressure and temperature T is given as:

$$e_i\left(p, T_i\right) = \frac{p + \gamma_i p_{\infty_i}}{p + p_{\infty_i}} c_{v_i} T_i.$$

Further, by following the procedure outlined in [68] for the Noble Abel Stiffened Gas EOS, an expression for the temperature T is derived as (see Appendix A.1):

$$T = \frac{1}{\rho c_v} \left[\frac{p + p_\infty - A\rho^\gamma}{\gamma - 1}\right].$$

Moreover, in the work by Le Métayer et al. [68], it is noted that allowing for A to be non-zero leads to isothermal curves that are non-monotonic i.e. for the same pressure and density, there exists two values for the temperature. To ensure that physical temperatures are computed, it is recommended in [68] to impose $A = 0$. Hence:

$$T = \frac{1}{\rho c_v} \left[\frac{p}{\gamma - 1} + \frac{p_\infty}{\gamma - 1}\right].$$

Finally, the speed of sound for each phase is expressed as:

$$c_i^2 = \gamma_i \frac{p + p_{\infty_i}}{\rho_i},$$

where the nomenclature has been previously defined. The derivation of the method to compute the mixture properties is detailed next.

2.4.3 Interface properties (mixed cells)

Consider the special case of an interface problem only involving constant pressure and velocity. Here, the momentum and energy equations reduce in 1-D to:

$$\frac{\partial \rho}{\partial t} + u \frac{\partial \rho}{\partial x} = 0, \tag{2.4a}$$

$$\frac{\partial}{\partial t}(\rho e) + u \frac{\partial}{\partial x}(\rho e) = 0. \tag{2.4b}$$

Substituting Equation (2.3) into Equation (2.4b) yields:

$$\frac{\partial}{\partial t}\left(\frac{p + \gamma p_\infty}{\gamma - 1}\right) + u \frac{\partial}{\partial x}\left(\frac{p + \gamma p_\infty}{\gamma - 1}\right) = 0.$$

Expanding the above:

$$\frac{\partial}{\partial t}\left(\frac{p}{\gamma - 1} + \frac{\gamma p_\infty}{\gamma - 1}\right) + u \frac{\partial}{\partial x}\left(\frac{p}{\gamma - 1} + \frac{\gamma p_\infty}{\gamma - 1}\right) = 0,$$

$$\frac{1}{\gamma - 1}\frac{\partial p}{\partial t} + p \frac{\partial}{\partial t}\left(\frac{1}{\gamma - 1}\right) + \frac{\partial}{\partial t}\left(\frac{\gamma p_\infty}{\gamma - 1}\right) + \frac{u}{\gamma - 1}\frac{\partial p}{\partial x} + pu \frac{\partial}{\partial x}\left(\frac{1}{\gamma - 1}\right) + u \frac{\partial}{\partial x}\left(\frac{\gamma p_\infty}{\gamma - 1}\right) = 0,$$

$$\frac{1}{\gamma - 1}\left[\frac{\partial p}{\partial t} + u \frac{\partial p}{\partial x}\right] + p \left[\frac{\partial}{\partial t}\left(\frac{1}{\gamma - 1}\right) + u \frac{\partial}{\partial x}\left(\frac{1}{\gamma - 1}\right)\right] +$$

$$\frac{\partial}{\partial t}\left(\frac{\gamma p_\infty}{\gamma - 1}\right) + u \frac{\partial}{\partial x}\left(\frac{\gamma p_\infty}{\gamma - 1}\right) = 0,$$

leads to:

$$p \left[\frac{\partial}{\partial t}\left(\frac{1}{\gamma - 1}\right) + u \frac{\partial}{\partial x}\left(\frac{1}{\gamma - 1}\right)\right] + \frac{\partial}{\partial t}\left(\frac{\gamma p_\infty}{\gamma - 1}\right) + u \frac{\partial}{\partial x}\left(\frac{\gamma p_\infty}{\gamma - 1}\right) = 0.$$

Since the solution is valid for any p, the following two equations must hold:

$$\frac{\partial}{\partial t}\left(\frac{1}{\gamma - 1}\right) + u \frac{\partial}{\partial x}\left(\frac{1}{\gamma - 1}\right) = 0, \tag{2.5a}$$

$$\frac{\partial}{\partial t}\left(\frac{\gamma p_\infty}{\gamma - 1}\right) + u \frac{\partial}{\partial x}\left(\frac{\gamma p_\infty}{\gamma - 1}\right) = 0. \tag{2.5b}$$

As was first seen in the work by Shyue [18] Equation (2.5) gives the specific form of the transport (advection) equations for the adiabatic properties to be satisfied in order to maintain a pressure equilibrium for the interface problem. Here, internal energy is expressed as:

$$\rho e = \rho_1 e_1 + \rho_2 e_2,$$
$$\frac{p}{\gamma - 1} + \frac{\gamma p_\infty}{\gamma - 1} = \alpha \frac{p}{\gamma_1 - 1} + \alpha \frac{\gamma_1 p_{\infty 1}}{\gamma_1 - 1} + (1 - \alpha) \frac{p}{\gamma_2 - 1} + (1 - \alpha) \frac{\gamma_2 p_{\infty 2}}{\gamma_2 - 1}.$$

For a constant pressure p, by splitting the above equation, expressions for computing the average properties at an interface (in mixed cells) are obtained:

$$\frac{1}{\gamma - 1} = \frac{\alpha}{\gamma_1 - 1} + \frac{1 - \alpha}{\gamma_2 - 1}, \tag{2.6a}$$

and,

$$\frac{\gamma p_\infty}{\gamma - 1} = \alpha \frac{\gamma_1 p_{\infty 1}}{\gamma_1 - 1} + (1 - \alpha) \frac{\gamma_2 p_{\infty 2}}{\gamma_2 - 1}. \tag{2.6b}$$

Moreover, by substituting the above into Equation (2.5), the 1-D VoF equation written in semi-conservative form is recovered as:

$$\frac{\partial \alpha}{\partial t} + \frac{\partial}{\partial x}(\alpha u) = \alpha \frac{\partial u}{\partial x}.$$

Intrinsically, the above demonstrates that consistently solving the classical VoF equation using the suitable expressions (Equation (2.6)) for the mixture properties enforces Equation (2.5) and hence guarantees a pressure equilibrium at the interface. In particular, the consistent numerical discretisation of this VoF and energy flux is a key topic of this book, which will be further expanded on in the next chapter.

Finally, the acoustic mixture velocity is computed via:

$$c^2 = \frac{p\left(\dfrac{1}{\gamma - 1} + 1\right) + \dfrac{\gamma p_\infty}{\gamma - 1}}{\rho \left(\dfrac{1}{\gamma - 1}\right)}. \tag{2.7}$$

The approximations presented in this section have been successfully used in the following numerical work [18, 24, 26, 32, 63].

2.5   Summary of governing equations

The governing equations for inviscid compressible flow written in compact strong form read as follows:

$$\frac{\partial U}{\partial t} + \nabla \cdot F\left(U\right) = S,\tag{2.8}$$

where F and S denote the flux and source term respectively with:

$$U = \begin{bmatrix} \rho \\ \rho u \\ \rho E \\ \alpha \end{bmatrix}, \qquad F\left(U\right) = \begin{bmatrix} \rho u \\ \rho u \otimes u + p I \\ \left(\rho E + p\right) u \\ \alpha u \end{bmatrix}, \qquad S = \begin{bmatrix} 0 \\ \sigma \kappa \nabla \alpha \\ \sigma \kappa \nabla \alpha \cdot u \\ \alpha \nabla \cdot u \end{bmatrix}.$$

The problem to be addressed is the general Initial Boundary Value Problem with the following characteristics:

1. Partial Differential Equation (2.8),

2. Initial Conditions: $U\left(\mathbf{x}, 0\right) = U^{(0)}\left(\mathbf{x}\right)$ with $\mathbf{x}$ denoting the position vector of a node in the domain,

3. Boundary Conditions: $U\left(\mathbf{x}_b, t\right) = U_b$, where subscript b denotes a boundary node.

Finally, the Initial Value Problem (IVP) to be solved is defined as:

$$U\left(\mathbf{x}, 0\right) = \begin{cases} U_L, & \text{if } |\mathbf{x} - \mathbf{x}_r| \leq a_0, \\ U_R, & \text{if } |\mathbf{x} - \mathbf{x}_r| > a_0, \end{cases}$$

where $\mathbf{x}_r$ denotes a reference coordinate, a_0 is the locus of points defining the region at a given state within the domain with subscripts $\{L, R\}$ denoting left and right states respectively.

Chapter 3

Numerical Methodology

3.1 Introduction

Over the last three decades, a number of Riemann type solvers have been employed for multi-phase compressible flow [1, 18, 26, 32, 48, 63]. For example, Shyue [18, 48] implemented the well-known class of Roe solvers [69] for both the void volume fraction and the gamma based model. However, this solver is known to be entropy violating due to the fact that it misinterprets a rarefaction as a shock wave leading to a rarefaction shock [20]. To address this, an entropy fix is proposed in the cited article and is not explored in this work.

A second class of popular Riemann type solvers are those proposed by Hartens, Lax, and Leer [70] (HLL). In their work, two wave configurations were used to obtain an inter-cell flux directly. However, their formulation did not account for the intermediate wave-speed which lead to the loss of contact discontinuities. To address this, Toro et al. [21] presented the HLLC approximate Riemann solver where the missing contact wave was restored. In particular, the main characteristic of HLLC is the ability to capture propagating sonic waves sharply ensuring the positivity of conserved variables [20, 71].

Further, higher order spatial reconstruction methods such as the Monotone Upstream-centred Scheme for Conservation Laws (MUSCL) [2], the Essentially Non-Oscillatory (ENO) or Weighted Essentially Non-Oscillatory (WENO) [72] scheme have traditionally been employed for the interpolation of conserved variables. However, as discussed in Chapter 1 and demonstrated in Chapter 4, when applied to the VoF field, these methods lead to the smearing of the liquid-gas interface and therefore inaccuracies in mixed cell properties (e.g. energy content and surface tension).

In recent years, extensive work [1, 26, 30, 32] has been done with respect to HLLC to circumvent the inherent diffusivity at the interface. In particular, different interface handling methods and modifications to the solver have been proposed. These have been critiqued in Chapter 1. Yet, out of the aforementioned articles, it is worth noting that only [1, 30] include surface tension effects. Moreover, to the author's knowledge, only Jibben et al. [30] take into account liquid availability in mixed cells (see Section 3.5) for the advection of the liquid-gas interface while using conserved variables for interface fluxing (as opposed to this work which advocates the use of primitive variables).

Hence, the goal of this chapter is to combine the volume conservative properties of a VoF method and the sharp sonic modelling attributes of the HLLC-MUSCL Riemann type solver [1, 2, 21]. Notably, the proposed method is based on the work by Garrick et al. [1]. However, in contrast to the latter cited article, the interface handling scheme is the algebraic donor-acceptor VoF CICSAM [35] method where the reconstruction of primitive variables allows for the consistent integration of the energy and VoF flux term.

The chapter is organised as follows. First, the discretisation of the spatial and temporal terms are detailed. This is followed by the derivation of the numerical face-flux reconstructor used to obtain second-order spatial accuracy on the primitive and conserved variables. Finally, the solvers employed to compute the inter-cell flux are described.

3.2   Discretisation of spatial and temporal terms

3.2.1   FV Median Dual Cell mesh Variant

In this work, the chosen discretisation method is the FV vertex centred median dual cell mesh variant since it is applicable to general meshes. Consider a node l connected to a node m where $m \in \{1, 2, ..., N_c\}$ with N_c denoting the number of neighbouring connected nodes as shown schematically in Figure 3.1 for an unstructured mesh. A computational median dual cell, Ω_l, is constructed around node l by connecting the mid-points and centroids of the adjacent edges and elements respectively. Consequently, the constructed control volume of capacity $\mathcal{V}_l$ is made up of sub-volumes bounded by a set of faces, denoted by $\partial\Omega_l$ with $\partial\Omega_l = \{\partial\Omega_{l,m_1}, \partial\Omega_{l,m_2}, ...\}$ and of N_b outer boundary faces of the set $A_b = \{A_{b_{f,1}}, A_{b_{f,2}}, ...\}$.

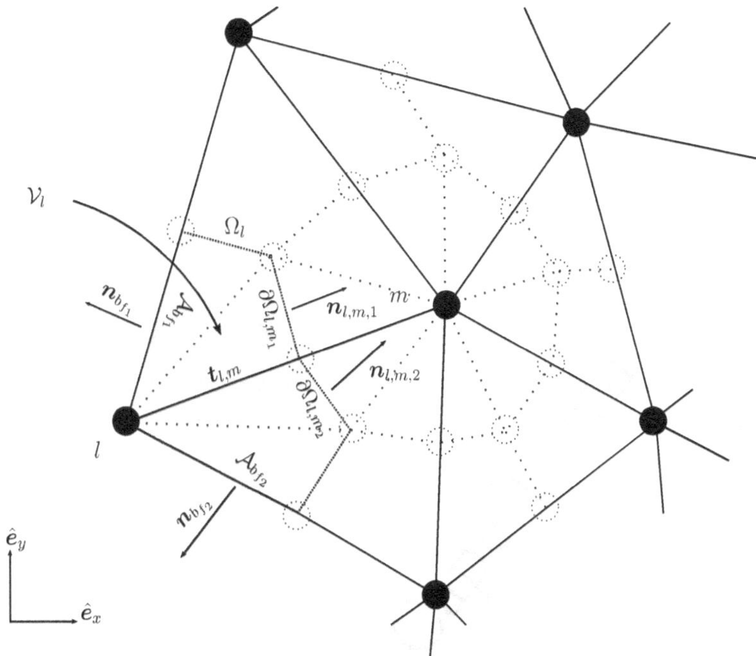

Figure 3.1: 2-D Computational median dual-cell around node l.

For an edge connecting nodes l to m, denoted $t_{l,m}$ of length $|t_{l,m}| = |\mathbf{x}_l - \mathbf{x}_m|$, the face coefficient is given by the sum of the product of the area and the normals of the shared segments which straddle the edge. In 2-D, this reads:

$$\mathcal{A}n|_{\partial\Omega_{l,m}} = \mathcal{A}_{\partial\Omega_{l,m_1}}\, n_{l,m_1} + \mathcal{A}_{\partial\Omega_{l,m_2}}\, n_{l,m_2}.$$

Similarly for a boundary face, the coefficient is computed as:

$$\mathcal{A}n|_{b_f} = \mathcal{A}_{b_{f_1}}\, n_{b_{f_1}} + \mathcal{A}_{b_{f_2}}\, n_{b_{f_2}}.$$

With the above geometric constructions set up in ELEMENTAL® , the discretisation is done in an edge-wise manner. The net flux through the computational cell Ω_l is therefore the sum of the individual fluxes across each internal face, f, and outer boundary face, b_f.

For Ω_l, this reads:

$$\frac{\partial \boldsymbol{U}_l}{\partial t} + \frac{1}{\mathcal{V}_l} \sum_{f \in \partial \Omega_l} (\boldsymbol{F} \boldsymbol{n} \mathcal{A}) \mid_f + \frac{1}{\mathcal{V}_l} \sum_{b_f \in \mathcal{A}_b} (\boldsymbol{F} \boldsymbol{n} \mathcal{A}) \mid_{b_f} = \boldsymbol{S}_l.$$

For the purpose of computing the face-flux values, HLLC is employed. The involved left and right states are reconstructed via a second-order accurate method. This is detailed in the next section.

3.2.2   Second-order spatial Reconstruction

The Monotone Upstream-centred Scheme for Conservation Laws [2] (MUSCL) is implemented in this work to obtain second-order accuracy on smooth fields for the left and the right state. Consider the edge $\boldsymbol{t}_{l,m}$ where the projected left and right nodes are denoted by l^* and m^* respectively as illustrated in Figure 3.2 (for internal edges). For an arbitrary variable ϕ, the gradient, $\nabla \phi$, at the node l, is first computed using central differencing as per the FV approach written as:

$$\nabla \phi \mid_l \approx \frac{1}{\mathcal{V}_l} \sum_{f \in \partial \Omega_l} (\phi \boldsymbol{n} \mathcal{A}) \mid_f,$$

where $\phi_f = \frac{\phi_l + \phi_m}{2}$.

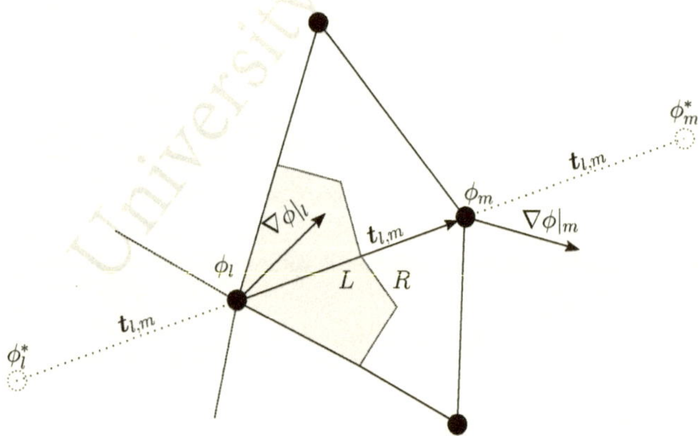

Figure 3.2: Projected up-wind and down-wind node for an edge.

The dot product of this gradient with the unit vector tangent to the edge gives the left-biased gradient in ϕ along the direction $\boldsymbol{t}_{l,m}$:

$$\nabla_{l,m}\phi|_l = \nabla\phi|_l \cdot \frac{\boldsymbol{t}_{l,m}}{|\boldsymbol{t}_{l,m}|}. \tag{3.1}$$

With respect to the projected left node l^*, along the edge, the above gradient can also be approximated using central differencing as:

$$\nabla_{l,m}\phi|_l \approx \frac{\phi_m - \phi_l^*}{2|\boldsymbol{t}_{l,m}|}. \tag{3.2}$$

Using Equations (3.1) and (3.2), the predicted ϕ value at node l^* is then computed as:

$$\phi_l^* = \phi_m - 2\nabla\phi|_l \cdot \boldsymbol{t}_{l,m}, \tag{3.3}$$

Similarly, ϕ_m^* is expressed as:

$$\phi_m^* = \phi_l + 2\nabla\phi|_m \cdot \boldsymbol{t}_{l,m}. \tag{3.4}$$

Further, adding and subtracting ϕ_l to Equation (3.3) yields:

$$\phi_m - \phi_l + \phi_l - \phi_l^* = 2\nabla\phi|_l \cdot \boldsymbol{t}_{l,m}.$$

From which two differences are then defined. First, the difference in ϕ between node l and m as:

$$\Delta\phi = \phi_m - \phi_l.$$

Second, the difference in the left state as:

$$\Delta\phi_L = \phi_l - \phi_l^*.$$

Using the above equations, an expression for the change in the left state is obtained:

$$\Delta\phi_L = 2\nabla\phi|_l \cdot \boldsymbol{t}_{l,m} - \Delta\phi.$$

Similarly for the difference in the right state, by defining $\Delta\phi_R = \phi_m^* - \phi_m$ and substituting into Equation (3.4) yields:

$$\Delta\phi_R = 2\nabla\phi|_m \cdot \boldsymbol{t}_{l,m} - \Delta\phi.$$

The above expressions allow for the computation of the left and right states (ϕ_L and ϕ_R) via a limiter, which is required due to the presence of sonic shocks:

$$\phi_L = \phi_l + \frac{1}{2}\psi\left(\frac{\Delta\phi_L}{\Delta\phi}\right)\Delta\phi, \qquad \text{and} \qquad \phi_R = \phi_m - \frac{1}{2}\psi\left(\frac{\Delta\phi_R}{\Delta\phi}\right)\Delta\phi. \tag{3.5}$$

Here, $\psi(r)$ is the flux limiter with $r = \frac{\Delta\phi_k}{\Delta\phi}$ and k denotes an arbitrary side $k \in \{L, R\}$. Finally, the van Albada [73] flux limiter is implemented in this work as follows:

$$\psi(r) = \begin{cases} \frac{r(r+1)}{r^2+1}, & \text{if } r \geq 0 \\ 0, & \text{otherwise} \end{cases}$$

where the nomenclature has been previously defined.

3.2.3 Temporal Integration and stability

In this work, the time integration scheme is the explicit second-order Runge-Kutta method defined as:

$$\boldsymbol{U}_l^{n+\frac{1}{2}} = \boldsymbol{U}_l^n - \frac{\Delta t}{2}\left[\nabla \cdot \boldsymbol{F}(\boldsymbol{U}_l^n) - \boldsymbol{S}_l^n\right], \tag{3.6a}$$

$$\boldsymbol{U}_l^{n+1} = \boldsymbol{U}_l^n - \Delta t\left[\nabla \cdot \boldsymbol{F}\left(\boldsymbol{U}_l^{n+\frac{1}{2}}\right) - \boldsymbol{S}_l^{n+\frac{1}{2}}\right]. \tag{3.6b}$$

Here, n is time and Δt denotes the global stable time step-size which is computed as:

$$\Delta t = \min\{\Delta t_c, \Delta t_\sigma\}. \tag{3.7}$$

For inviscid compressible flow:

$$\Delta t_c = \text{CFL} \min_{l \in N} \frac{\Delta \mathsf{x}_l}{|\boldsymbol{u}_l| + c_l},$$

with N denoting the total number of nodes in the domain, c_l is computed as per Equation (2.7), $\Delta\mathsf{x}_l$ denotes the effective mesh spacing for a node l. Further, CFL denotes the Courant-Frederichs-Lewy number where $\text{CFL} \in (0,1)$. Finally, the time constraint for surface tension effects is given as per Brackbill et al. [5]:

$$\Delta t_\sigma = \min_{l \in N} \sqrt{\frac{(\rho_1 + \rho_2)\, \Delta x_l^3}{4\pi\sigma}},$$

where the nomenclature has been previously defined.

3.2.4 Eigenvalues and vectors

The solution to the Riemann problem consists of a family of waves associated with a set of eigenvalues for the given system of hyperbolic governing equations. Figure 3.3 illustrates an example of such a solution in an x-t plane, which includes three waves. Firstly, a rarefaction wave which can be described as a fan-based type wave where there exists continuous or linear variations in physical quantities. Secondly, a shock wave, where variations are characterised by discontinuous jumps. Thirdly, the contact wave, where in the absence of surface tension, there is no jump in pressure nor particle velocity. Naturally, it follows that in order to account for surface tension effects, the Laplace pressure condition must be enforced at the contact wave and will be further expanded on in Section 3.3.

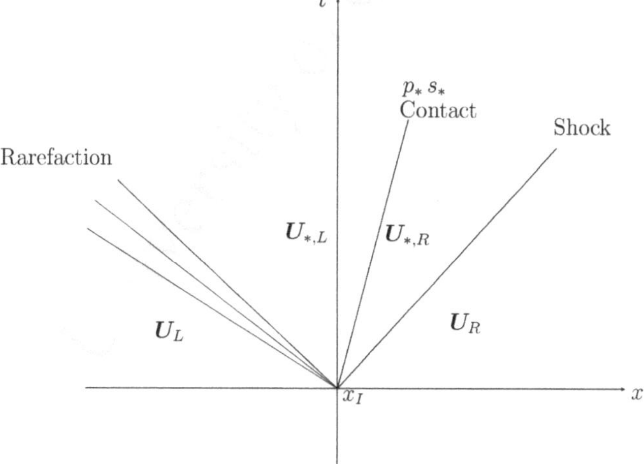

Figure 3.3: Elementary wave configuration to the Riemann Problem.

Each eigenvalue and associated vector carries information that will propagate across the domain. In particular, the solution must satisfy the Rankine-Hugoniot jump condition which states that the change in flux is the product of the wave-speed and jump in U as:

$$\Delta F = s\Delta U,$$

where s is the wave-speed of a discontinuous wave solution (left, right or contact discontinuity). The above forms the basis for the derivation of the inter-cell HLLC flux. Next, the eigenvalues and vectors for the system of governing equations are derived.

First, the governing equation set is expressed in strong compact form in 2-D as:

$$\frac{\partial U}{\partial t} + \frac{\partial F_x}{\partial x} + \frac{\partial F_y}{\partial y} = S,$$

where subscript x and y denote the flux in the respective Cartesian direction. Further

$$F_x(U) = \begin{bmatrix} \rho u \\ \rho u^2 + p \\ \rho uv \\ (\rho E + p)\,u \\ \alpha u \end{bmatrix}, \quad F_y(U) = \begin{bmatrix} \rho v \\ \rho uv \\ \rho v^2 + p \\ (\rho E + p)\,v \\ \alpha v \end{bmatrix}, \quad \text{and} \quad S = \begin{bmatrix} 0 \\ \sigma\kappa\frac{\partial\alpha}{\partial x} \\ \sigma\kappa\frac{\partial\alpha}{\partial y} \\ \sigma\kappa\left(u\frac{\partial\alpha}{\partial x} + v\frac{\partial\alpha}{\partial y}\right) \\ \alpha\left(\frac{\partial u}{\partial x} + \frac{\partial v}{\partial y}\right) \end{bmatrix}.$$

The above system of equations may be written in terms of the primitive variables $W = \begin{bmatrix} \rho, u, v, p, \alpha \end{bmatrix}^T$ in linear quasi-conservative form as:

$$\frac{\partial W}{\partial t} + A(W)\frac{\partial W}{\partial x} + B(W)\frac{\partial W}{\partial y} = 0,$$

where $A(W)$ and $B(W)$ are the Jacobian matrices associated with x and y direction, respectively. Specifically:

$$A(W) = \begin{bmatrix} u & \rho & 0 & 0 & 0 \\ 0 & u & 0 & \frac{1}{\rho} & \frac{-\sigma\kappa}{\rho} \\ 0 & 0 & u & 0 & 0 \\ 0 & \rho c^2 & 0 & u & 0 \\ 0 & 0 & 0 & 0 & u \end{bmatrix}.$$

The eigenvalues associated with the Jacobian matrix $\boldsymbol{A}$ are:

$$\boldsymbol{\Lambda_A} = diag(u - c, u, u, u, u + c),$$

with *diag* denoting a diagonal matrix. The corresponding eigenvectors are:

$$\boldsymbol{K_A} = \begin{bmatrix} \rho & 1 & 0 & 0 & \rho \\ -c & 0 & 1 & 0 & c \\ 0 & 0 & 0 & 0 & 0 \\ \rho c^2 & 0 & 0 & \sigma\kappa & \rho c^2 \\ 0 & 0 & 0 & 1 & 0 \end{bmatrix}.$$

The characteristic eigenvector corresponding to the Jacobian $\boldsymbol{B}$ matrix associated with the remaining dimension can be easily derived in a similar way and is not included here. Finally, for a system of equations of arbitrary dimension d, normal to the interface, there are three unique and real eigenvalues; $\lambda_1 = \boldsymbol{u} \cdot \boldsymbol{n} - c$, $\lambda_2 = \boldsymbol{u} \cdot \boldsymbol{n}$ and $\lambda_3 = \boldsymbol{u} \cdot \boldsymbol{n} + c$ (multiplicity of spatial dimensions + one for each transport equation), where the interface normal is $\boldsymbol{n}$.

Notably, the characteristic field associated with $\boldsymbol{K}^{(2)}$ is linearly degenerate i.e. $\nabla\lambda_2 \cdot \boldsymbol{K}^{(2)} = 0$ and therefore denotes a contact discontinuity, across which consistency conditions will be imposed while the fields associated with $\boldsymbol{K}^{(1)}$ and $\boldsymbol{K}^{(3)}$ are strongly non-linear. These can either be shock waves (discontinuous) or rarefaction waves (linear fan type).

3.3   Multi-dimensional two-phase HLLC solver

The elementary wave solution for HLLC consists of three waves (s_L, s_* and s_R) separating four states namely, $\boldsymbol{U}_L$, $\boldsymbol{U}_{*L}$, $\boldsymbol{U}_{*R}$ and $\boldsymbol{U}_R$ with subscript $*$ denoting the intermediate star-state associated with the contact wave as shown in Figure 3.3. Intrinsically, each region (left, right and intermediate) is defined as per the characteristic wave-speeds where the inter-cell flux, $\boldsymbol{F}_f$, is derived from the Rankine-Hugoniot jump conditions as:

$$\boldsymbol{F}_f = \begin{cases} \boldsymbol{F}_L, & \text{if } \ 0 \leq s_L, \\ \boldsymbol{F}_L + s_L \left(\boldsymbol{U}_{*,L} - \boldsymbol{U}_L \right), & \text{if } \ s_L \leq 0 \leq s_*, \\ \boldsymbol{F}_R + s_R \left(\boldsymbol{U}_{*,R} - \boldsymbol{U}_R \right), & \text{if } \ s_* \leq 0 \leq s_R, \\ \boldsymbol{F}_R, & \text{if } \ 0 \geq s_R. \end{cases} \tag{3.8}$$

Here, Equation (3.8) gives the generic form for all existing HLLC solvers in literature. This defines the starting point for the inclusion of CICSAM and surface tension effects in the solver. In particular, the improvements that will be proposed for the computation of the intermediate flux will draw from existing work on HLLC [1] and will be further expanded on in the next sections.

3.3.1   Consistency conditions

To obtain the intermediate star-state flux, consistency conditions on the velocity and pressure fields must first be set. For the face normal and tangential components of the velocity field at the contact discontinuity, the following must hold:

$$\boldsymbol{u}_{*,L} \cdot \boldsymbol{n} = \boldsymbol{u}_{*,R} \cdot \boldsymbol{n} = s_* \quad \text{and} \quad \boldsymbol{u}_{*k} - \left(\boldsymbol{u}_{*k} \cdot \boldsymbol{n} \right) \boldsymbol{n} = \boldsymbol{u}_k - \left(\boldsymbol{u}_k \cdot \boldsymbol{n} \right) \boldsymbol{n}, \tag{3.9}$$

with $k \in \{L, R\}$. Traditionally, in the absence of surface tension, the following pressure condition is enforced:

$$p_{*,L} = p_{*,R} = p_*.$$

However, to account for the pressure jump induced by the interface curvature, a new consistency condition must be derived. Using the generalised Riemann invariants and eigenvector $\boldsymbol{k}^{(4)} = \begin{bmatrix} 0, 0, 0, \sigma\kappa, 1 \end{bmatrix}^T$, the Young-Laplace pressure jump condition is obtained:

$$\Delta p_* = \sigma\kappa\Delta\alpha,$$
$$p_{*,R} - p_{*,L} = \sigma\kappa \left(\alpha_R - \alpha_L \right), \tag{3.10}$$

where this consistency pressure condition was first used in [1].

3.3.2 Inviscid HLLC-CICSAM edge flux

Numerical Energy Consistency Criteria

The incorporation of CICSAM into HLLC begins by noting a key difference between the manner in which MUSCL and CICSAM operate. Notably, MUSCL is a monotone up-winding scheme which interpolates left and right states on either side of a face. These are then blended by HLLC based on characteristic wave-speeds to arrive at a discretised face-flux. In contrast, CICSAM employs a blend of up-wind differencing and down-winding to reconstruct a flux based on availability (amount of fluid available in the cell). As a result, CICSAM yields a more accurate discretisation of the α flux in the VoF equation. As HLLC is more desirable where compressible flow characteristics are key ($\rho, \boldsymbol{u}, p$ etc.), a consistent blend with CICSAM is sought. To enable this, face-fluxes are discretised in terms of primitive variables ($\rho, \boldsymbol{u}, p$) in this work. This is illustrated next.

Consider again the special 1-D interface propagation case described in Section 2.4.3 where the system reduces to the continuity and thermal energy equations written in conservative form as:

$$\frac{\partial \rho}{\partial t} + \frac{\partial}{\partial x}\left(\rho u\right) = 0, \tag{3.11a}$$

$$\frac{\partial}{\partial t}\left(\rho e\right) + \frac{\partial}{\partial x}\left(\rho e u\right) = 0. \tag{3.11b}$$

The VoF equation is added:

$$\frac{\partial \alpha}{\partial t} + \frac{\partial}{\partial x}\left(\alpha u\right) = 0. \tag{3.11c}$$

Since the sharp propagation of the interface is desired, the semi-discrete form of the VoF equation reads:

$$\frac{\partial \alpha}{\partial t} \approx -\frac{1}{\mathcal{V}_l} \sum_{f \in \partial \Omega_l} \alpha_{f,C} u_{ff}, \tag{3.12}$$

where $\alpha_{f,C}$ denotes the face value associated with CICSAM and u_{ff} is the face-flux given by Equation (3.20). If the energy field is reconstructed using the HLLC-MUSCL approach, the thermal equation is discretised as:

$$\frac{\partial}{\partial t}\left(\rho e\right) \approx -\frac{1}{\mathcal{V}_l} \sum_{f \in \partial \Omega_l} \left(\rho e\right)_{f,H} u_{ff},$$

where subscript H denotes the internal energy flux obtained from HLLC associated with the MUSCL reconstruction. Introducing the Gas Stiffened EOS, the above expression follows as:

$$\frac{\partial}{\partial t}\left(\frac{p+\gamma p_\infty}{\gamma-1}\right) \approx -\frac{1}{\mathcal{V}_l}\sum_{f\in\partial\Omega_l}\left(\frac{p+\gamma p_\infty}{\gamma-1}\right)_{f,H} u_{ff}.$$

The above equation can be split into two transport sub-equations as:

$$\frac{\partial}{\partial t}\left(\frac{p}{\gamma-1}\right) \approx -\frac{1}{\mathcal{V}_l}\sum_{f\in\partial\Omega_l}\left(\frac{p}{\gamma-1}\right)_{f,H} u_{ff}, \tag{3.13a}$$

$$\frac{\partial}{\partial t}\left(\frac{\gamma p_\infty}{\gamma-1}\right) \approx -\frac{1}{\mathcal{V}_l}\sum_{f\in\partial\Omega_l}\left(\frac{\gamma p_\infty}{\gamma-1}\right)_{f,H} u_{ff}. \tag{3.13b}$$

Since the pressure is constant, Equation (3.13a) reduces to:

$$\frac{\partial}{\partial t}\left(\frac{1}{\gamma-1}\right) \approx -\frac{1}{\mathcal{V}_l}\sum_{f\in\partial\Omega_l}\left(\frac{1}{\gamma-1}\right)_{f,H} u_{ff},$$

which can be further expanded in terms of the volume fraction field as:

$$\frac{\partial}{\partial t}\left(\frac{\alpha}{\gamma_1-1}+\frac{1-\alpha}{\gamma_2-1}\right) \approx -\frac{1}{\mathcal{V}_l}\sum_{f\in\partial\Omega_l}\left(\frac{\alpha_{f,H}}{\gamma_1-1}+\frac{1-\alpha_{f,H}}{\gamma_2-1}\right) u_{ff}, \tag{3.14}$$

where $\alpha_{f,H}$ represents the face volume fraction field that is consistent with the HLLC-MUSCL reconstruction of the energy field. Equation (3.14) can be summarised for a phase $i \in \{1,2\}$ as:

$$\frac{\partial}{\partial t}\left(\frac{\alpha_i}{\gamma_i-1}\right) \approx -\frac{1}{\mathcal{V}_l}\sum_{f\in\partial\Omega_l}\frac{\alpha_{i_{f,H}}}{\gamma_i-1} u_{ff},$$

where the same conclusion can drawn from Equation (3.13b). Further simplification of the above expression leads to:

$$\frac{\partial\alpha_i}{\partial t} \approx -\frac{1}{\mathcal{V}_l}\sum_{f\in\partial\Omega_l}\alpha_{i_{f,H}} u_{ff}, \tag{3.15}$$

where $\alpha_i \in \{\alpha, 1-\alpha\}$ for a two-phase flow. It is clear that Equation (3.15) is inconsistent with Equation (3.12) as $\alpha_{f,C} \neq \alpha_{f,H}$ which will lead to the appearance of spurious currents in the pressure as well as the velocity field.

The above demonstrates that a consistent discretisation method must be employed on both the energy and VoF flux in order to maintain an oscillation free field. This relationship is set up by the caloric Gas Stiffened EOS and will be referred to as the *Energy Consistency Criteria* (ECC). Hence, with the introduction of the VoF method, the ECC can only be achieved if the energy field is discretised via the EOS by applying a blend of HLLC and CICSAM as follows:

$$\frac{\partial}{\partial t}\left(\rho e\right) \approx -\frac{1}{\mathcal{V}_l}\sum_{f\in\partial\Omega_l}\left(\rho e\right)_f u_{ff} = \sum_{f\in\partial\Omega_l}\left(p_{f,H}\left.\frac{1}{\gamma-1}\right|_{f,C} + \left.\frac{\gamma p_\infty}{\gamma-1}\right|_{f,C}\right)u_{ff}.$$

Expanding the above as per Equations (3.13) and (3.14) will lead to the desired VoF discretisation, i.e. Equation (3.12) is recovered. This ensures consistency between the VoF CICSAM flux and the energy term computed by HLLC. Further, since at the face of an edge, sharp interface properties are sought, it is proposed in this work that the EOS properties are discretised using CICSAM as follows:

$$\left.\frac{1}{\gamma-1}\right|_{f,C} = \frac{\alpha_{f,C}}{\gamma_1-1} + \frac{1-\alpha_{f,C}}{\gamma_2-1}, \quad \text{and,} \quad \left.\frac{\gamma p_\infty}{\gamma-1}\right|_{f,C} = \frac{\alpha_{f,C}\gamma_1 p_{\infty_1}}{\gamma_1-1} + \frac{\alpha_{f,C}\gamma_2 p_{\infty_2}}{\gamma_2-1}. \quad (3.16)$$

This ensures an accurate discretisation of the energy field as will be shown in Chapter 4.

Intermediate star-state

In this sub-section, the following notation will be employed to differentiate between different scheme variants in discretising a variable ϕ at a face: $\phi_{k,M}$ will refer to MUSCL, $\phi_{f,C}$ to CICSAM and ϕ_k on its own will define a combination of the two aforementioned operators for an arbitrary side $k \in \{L, R\}$. The HLLC normal flux can be therefore re-written from Equation (3.8) as:

$$\boldsymbol{F}_f = \frac{1+\text{sign}\left(s_*\right)}{2}\left[\boldsymbol{F}_L + s^-\left(\boldsymbol{U}_{*,L} - \boldsymbol{U}_L\right)\right] + \frac{1-\text{sign}\left(s_*\right)}{2}\left[\boldsymbol{F}_R + s^+\left(\boldsymbol{U}_{*,R} - \boldsymbol{U}_R\right)\right], \quad (3.17)$$

where the flux for an arbitrary side k is computed via:

$$\boldsymbol{F}_k = \begin{bmatrix} \rho_{k,M}\boldsymbol{u}_{k,M}\cdot\boldsymbol{n} \\ \rho_{k,M}\boldsymbol{u}_{k,M}\boldsymbol{u}_{k,M}\cdot\boldsymbol{n} + p_{k,M}\boldsymbol{n} \\ \left(\rho_{k,M}E_k + p_{k,M}\right)\boldsymbol{u}_{k,M}\cdot\boldsymbol{n} \\ \alpha_{f,C}\boldsymbol{u}_{k,M}\cdot\boldsymbol{n} \end{bmatrix},$$

and the intermediate star-state is derived as shown in Appendix A.2 as:

$$\boldsymbol{U}_{*_k} = \chi_k \begin{bmatrix} \rho_{k,M} \\ \rho_{k,M}\boldsymbol{u}_{k,M} + \rho_{k,M}\left(s_* - \boldsymbol{u}_{k,M}\cdot\boldsymbol{n}\right)\boldsymbol{n} \\ \rho_{k,M}E_k + \rho_{k,M}\left(s_* - \boldsymbol{u}_{k,M}\cdot\boldsymbol{n}\right)\left(s_* + \dfrac{p_{k,M}}{\rho_{k,M}\left(s_k - \boldsymbol{u}_{k,M}\cdot\boldsymbol{n}\right)}\right) \\ \alpha_{f,C} \end{bmatrix}.$$

The sonic characteristic, χ_k, is defined as:

$$\chi_k = \frac{s_k - \boldsymbol{u}_{k,M}\cdot\boldsymbol{n}}{s_k - s_*},$$

where s_* denotes the contact wave computed as per [1, 71]:

$$s_* = \underbrace{\frac{p_{L,M} - p_{R,M} + \rho_{R,M}\boldsymbol{u}_{R,M}\cdot\boldsymbol{n}\left(s_R - \boldsymbol{u}_{R,M}\cdot\boldsymbol{n}\right) - \rho_{L,M}\boldsymbol{u}_{L,M}\cdot\boldsymbol{n}\left(s_L - \boldsymbol{u}_{L,M}\cdot\boldsymbol{n}\right)}{\rho_{R,M}\left(s_R - \boldsymbol{u}_{R,M}\cdot\boldsymbol{n}\right) - \rho_{L,M}\left(s_L - \boldsymbol{u}_{L,M}\cdot\boldsymbol{n}\right)}}_{\text{Convective flux term}}$$
$$- \underbrace{\frac{\sigma\kappa\left(\alpha_{L,\sigma} - \alpha_{R,\sigma}\right)}{\rho_{R,M}\left(s_R - \boldsymbol{u}_{R,M}\cdot\boldsymbol{n}\right) - \rho_{L,M}\left(s_L - \boldsymbol{u}_{L,M}\cdot\boldsymbol{n}\right)}}_{\text{Surface tension term}}.$$

$$(3.18)$$

where in the above, $\alpha_{k,\sigma}$ denotes the face value of alpha which is consistent with the discretisation of the surface tension source term and will be further expanded on in Section 3.3.3.

The energy field is discretised as:

$$(\rho E)_k = \frac{1}{2}\rho_{k,M}\boldsymbol{u}_{k,M}\cdot\boldsymbol{u}_{k,M} + (\rho e)_k\,, \qquad (3.19\text{a})$$

where

$$(\rho e)_k = p_{k,M}\frac{1}{\gamma - 1}\bigg|_{f,C} + \frac{\gamma p_\infty}{\gamma - 1}\bigg|_{f,C}. \qquad (3.19\text{b})$$

The wave-speeds are computed as:

$$s^- = \min(0, s_L) \quad \text{and} \quad s^+ = \max(0, s_R),$$

where s_L and s_R denote the characteristic left and right wave-speed respectively. As per

Einfeldt et al. [74], the wave-speeds can be estimated using the Roe average eigenvalues as:

$$s_L = \min\left(\tilde{\boldsymbol{u}} \cdot \boldsymbol{n} - \tilde{c}, \boldsymbol{u}_{L,M} \cdot \boldsymbol{n} - c_L\right) \quad \text{and,} \quad s_R = \max\left(\tilde{\boldsymbol{u}} \cdot \boldsymbol{n} + \tilde{c}, \boldsymbol{u}_{R,M} \cdot \boldsymbol{n} + c_R\right),$$

where

$$c_k = \sqrt{\frac{1}{\frac{1}{\gamma-1}|_{f,C}}\left(H_k - \frac{1}{2}\boldsymbol{u}_{k,M} \cdot \boldsymbol{u}_{k,M}\right)}.$$

Further, H_k is the specific stagnation enthalpy computed via:

$$H_k = E_k + \frac{p_{k,M}}{\rho_{k,M}}.$$

Finally, $\tilde{\phi} \in \{\tilde{u}, \tilde{H}\}$ is computed using:

$$\tilde{\phi} = \frac{\sqrt{\rho_L}\phi_L + \sqrt{\rho_R}\phi_R}{\sqrt{\rho_L} + \sqrt{\rho_R}}.$$

where the nomenclature has been previously defined.

Face-flux

Given the equation for the intermediate flux (Equation (3.17)), a mathematical equivalent expression for the face-flux, u_{ff}, that is consistent with the proposed CICSAM discretisation can be derived. Using an adaptation to the HLLC solver proposed by Johnsen et al. [26], it is assumed that $(\alpha u_{ff}) = \alpha_{f,C} u_{ff}$. The expression for the alpha flux hence follows as:

$$(\alpha u_{ff}) = \frac{1 + \text{sign}(s_*)}{2}\left[\alpha_{f,C}\boldsymbol{u}_L \cdot \boldsymbol{n} + s^-\left(\chi_L \alpha_{f,C} - \alpha_{f,C}\right)\right]$$
$$+ \frac{1 - \text{sign}(s_*)}{2}\left[\alpha_{f,C}\boldsymbol{u}_R \cdot \boldsymbol{n} + s^+\left(\chi_R \alpha_{f,C} - \alpha_{f,C}\right)\right].$$

If $s_* > 0$, it follows that:

$$\alpha_{f,C}u_{ff} = \alpha_{f,C}\left(\boldsymbol{u}_L \cdot \boldsymbol{n} + s^-\left(\chi_L - 1\right)\right),$$
$$\implies u_{ff} = \boldsymbol{u}_L \cdot \boldsymbol{n} + s^-\left(\chi_L - 1\right).$$

Similarly, if $s_* < 0$:

$$\alpha_{f,C} u_{ff} = \alpha_{f,C}\left(\boldsymbol{u}_R \cdot \boldsymbol{n} + s^+ (\chi_R - 1)\right),$$
$$\implies u_{ff} = \boldsymbol{u}_R \cdot \boldsymbol{n} + s^+ (\chi_R - 1).$$

Hence, for the edge $\boldsymbol{t}_{l,m}$, the face-flux is:

$$u_{ff} = \left(\frac{1 + \mathrm{sign}\,(s_*)}{2}[\boldsymbol{u}_{L,M} \cdot \boldsymbol{n}_{l,m} s^- (\chi_L - 1)] + \frac{1 - \mathrm{sign}\,(s_*)}{2}[\boldsymbol{u}_{R,M} \cdot \boldsymbol{n}_{l,m} + s^+ (\chi_R - 1)]\right)\mathcal{A}_{l,m}.$$

$$(3.20)$$

where the nomenclature has been previously defined. In particular, the above definition allows for easy implementation of any VoF method with the HLLC solver in a manner that guarantees the contact-preserving properties.

Finally, though the VoF CICSAM interface reconstruction is used in this work, for the purpose of comparison, the alpha face value, $\alpha_{f,H}$, is also discretised using HLLC-MUSCL as:

$$\alpha_{f,H} = \begin{cases} \alpha_{L,M}, & \text{if } 0 \leq s_L, \\ \alpha_{*L} = \chi_L \alpha_{L,M}, & \text{if } s_L \leq 0 \leq s_*, \\ \alpha_{*R} = \chi_R \alpha_{R,M} & \text{if } s_* \leq 0 \leq s_R, \\ \alpha_{R,M}, & \text{if } 0 \geq s_R, \end{cases}$$

where the nomenclature has been previously defined.

3.3.3 Surface tension source terms

HLLC solvers rarely take into account surface tension effects for multi-phase compressible flow [1, 30]. The first mention of an HLLC based scheme with surface tension was by Garrick et al. [1] where a compression technique is used to remove any numerical diffusion present in the VoF field. In contrast, in this work, the conservative VoF-CICSAM method is combined with the HLLC solver to allow for curvature to be computed accurately. Similar to [1], this requires a well-balanced implementation with the surface tension source term. A well-balanced implementation implies that the surface tension source term is balanced by the computed pressure gradient in a steady flow field with zero velocity and zero gravity. This may be mathematically expressed by considering a static bubble configuration [5, 24, 75, 76] where the Euler equation reduces to:

$$-\nabla p + \sigma\kappa\nabla\alpha = 0.$$

The above is discretised for a node l as:

$$\frac{1}{\mathcal{V}_l}\sum_{f\in\partial\Omega_l}(p\boldsymbol{n}\mathcal{A})|_f+\varepsilon_p = \frac{\sigma\kappa_l}{\mathcal{V}_l}\sum_{f\in\partial\Omega_l}(\alpha\boldsymbol{n}\mathcal{A})|_f+\varepsilon_\sigma. \tag{3.21}$$

Here, ε_p and ε_σ are the discretisation errors associated with pressure and alpha respectively.

For the case where s^+ and s^- are non-zero, the pressure face as per Equation (3.17) is given as:

$$p_f = \frac{1+\text{sign}(s_*)}{2}\left[p_{L,M} + \chi_L s_L \rho_{L,M} s_*\right] + \frac{1-\text{sign}(s_*)}{2}\left[p_{R,M} + \chi_R s_R \rho_{R,M} s_*\right]. \tag{3.22}$$

Here, $\chi_k = \frac{s_k}{s_k-s_*}$, $s_L = \min(-\tilde{c}, -c_L)$ and $s_R = \max(\tilde{c}, c_R)$. The contact wave-speed simplifies to:

$$s_* = \frac{p_{L,M} - p_{R,M} - \sigma\kappa_f\left(\alpha_{L,\sigma} - \alpha_{R,\sigma}\right)}{\rho_{R,M}s_R - \rho_{L,M}s_L}. \tag{3.23}$$

In the case of a static bubble, as the pressure field converges to the analytical solution, the consistency pressure condition (Equation (3.10)) is satisfied and hence the contact wave-speed tends to zero (or machine precision, ϵ). Here, Equation (3.22) may therefore be expressed as:

$$p_f = \frac{1+\text{sign}(s_*)}{2}\left[p_{L,M} + \Upsilon_L\right] + \frac{1-\text{sign}(s_*)}{2}\left[p_{R,M} + \Upsilon_R\right],$$

where $\Upsilon_k \lll p_{k,M}$ since $\Upsilon_k \approx \epsilon$.

As explained by Popinet [76], to recover the discrete equilibrium of Equation (3.21), it is required for ε_p to cancel ε_σ to zero. To this end, the same operator used for the pressure face must also be employed for the volume fraction field in the surface tension term. Hence, in the interest of a consistent discretisation, $\alpha_{f,\sigma}$ is interpolated for the left or right state using MUSCL and up-winded using HLLC as follows:

$$\alpha_{f,\sigma} = \frac{1+\text{sign}(s_*)}{2}\alpha_{L,M} + \frac{1-\text{sign}(s_*)}{2}\alpha_{R,M},$$

Similarly, the contact wave-speed now reads:

$$s_* = \frac{p_{L,M} - p_{R,M} - \sigma\kappa_f\left(\alpha_{L,M} - \alpha_{R,M}\right)}{\rho_{R,M}\left(s_R - \boldsymbol{u}_{R,M}\cdot\boldsymbol{n}\right) - \rho_{L,M}\left(s_L - \boldsymbol{u}_{L,M}\cdot\boldsymbol{n}\right)},$$

where κ_f is central differenced as per [1] i.e. $\kappa_f = \frac{\kappa_l + \kappa_m}{2}$. This approach for the computation of κ_f is also widely used for solving the pressure Poiseuille equation in semi-implicit projection solvers [75, 76]. The approximation is valid for a sufficiently fine mesh and for cases where the curvature is uniform and constant.

Finally, the energy surface tension term is re-expressed in its conservative form and discretised as follows:

$$\frac{\sigma\kappa}{\mathcal{V}_l}\int_\Omega \nabla\alpha\cdot\boldsymbol{u}d\mathcal{V} \approx \frac{\sigma\kappa_l}{\mathcal{V}_l}\left(\sum_{f\in\partial\Omega_l}\alpha_{\sigma,f}u_{ff} - \alpha_l\sum_{f\in\partial\Omega_l}u_{ff}\right).$$

This guarantees a well-balanced method as demonstrated analytically in the next subsection and numerically via the static bubble test-case in Section 4.3.4.

3.3.4 Semi-Analytical Proof of Well-Balanced Scheme

Consider the special case where the interface lies half-way across an edge as shown for an equispaced mesh in in Figure 3.4. Here, the analytical pressure field is such that $p_1 - p_0 = \sigma\kappa_a$ and hence $s_* = 0$.

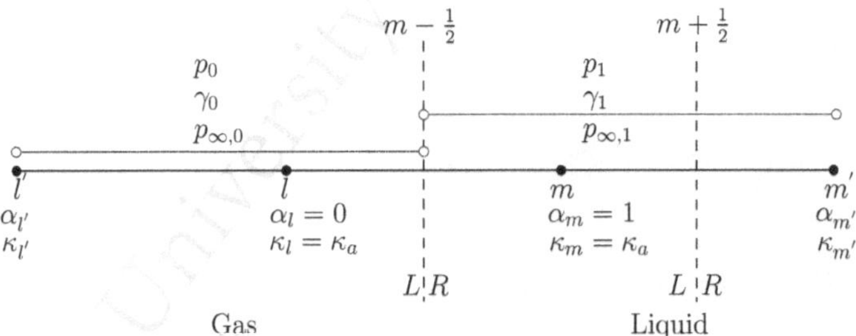

Figure 3.4: Well-balanced surface tension Disscretistion on equispaced mesh.

The momentum term at node m is discretised as per Equation (3.22):

$$\frac{1}{\mathcal{V}_m}\sum_{f\in\partial\Omega_m}(p\boldsymbol{n}\mathcal{A})\,|_f = \frac{1}{\mathcal{V}_m}\left[\frac{p_{m+\frac{1}{2}L} + p_{m+\frac{1}{2}R}}{2} - \frac{p_{m-\frac{1}{2}L} + p_{m-\frac{1}{2}R}}{2}\right]. \tag{3.24}$$

Here, interpolating using MUSCL, $p_{m+\frac{1}{2}_L} = p_{m-\frac{1}{2}_R} = p_m$, $p_{m+\frac{1}{2}_R} = p_{m'}$ and $p_{m-\frac{1}{2}_L} = p_l$. The momentum term now reads:

$$\frac{1}{\mathcal{V}_m} \sum_{f\in\partial\Omega_m} (p\boldsymbol{n}\mathcal{A})\,|_f = \frac{1}{2|\boldsymbol{t}_{l,m}|} \left[p_{m'} - p_l\right]. \tag{3.25}$$

Similarly, the surface tension term is discretised at node m as:

$$\frac{\sigma\kappa_m}{\mathcal{V}_m} \sum_{f\in\partial\Omega_m} (\alpha\boldsymbol{n}\mathcal{A})\,|_f = \frac{\sigma\kappa_m}{2|\boldsymbol{t}_{l,m}|} \left[\alpha_{m'} - \alpha_l\right] \tag{3.26}$$

Subtracting Equation (3.26) from Equation (3.25) yields:

$$\frac{1}{\mathcal{V}_m} \sum_{f\in\partial\Omega_m} (p\boldsymbol{n}\mathcal{A})\,|_f - \frac{\sigma\kappa_m}{\mathcal{V}_m} \sum_{f\in\partial\Omega_m} (\alpha\boldsymbol{n}\mathcal{A})\,|_f = \frac{1}{2|\boldsymbol{t}_{l,m}|} \left[p_{m'} - p_l - \sigma\kappa_m \left(\alpha_{m'} - \alpha_l\right)\right]. \tag{3.27}$$

Here, $p_{m'} = p_1$, $p_l = p_0$ and $\kappa_m = \kappa_a$. Substituting in Equation (3.27) and re-arranging yields:

$$p_1 - p_0 - \sigma\kappa_a\left(1 - 0\right) = \left[\sigma\kappa_a - \sigma\kappa_a\right] = 0. \tag{3.28}$$

This shows that the scheme is well-balanced.

3.4   Curvature Reconstruction

In this work, the curvature is re-constructed using two methods. Firstly, using the convolution method as proposed by Brackbill et al. [5] where the divergence of the normal of the filtered (smooth) alpha field α^* is used:

$$\kappa_l \approx -\nabla\cdot\hat{\boldsymbol{n}} \approx -\nabla\cdot\left.\frac{\nabla\alpha^*}{|\nabla\alpha^*|}\right|_l \approx -\frac{1}{\mathcal{V}_l} \sum_{f\in\partial\Omega_l} \left.\left(\frac{\nabla\alpha^*}{|\nabla\alpha^*|}\cdot\boldsymbol{n}\mathcal{A}\right)\right|_f.$$

The above is in the interest of applicability to arbitrary meshes in this work. Secondly, using height functions [3,4], which were recently added to ELEMENTAL® to obtain second-order spatial accuracy on Cartesian meshes.

3.5   VoF equation

The final equation to be discretised is the VoF equation:

$$\frac{1}{\mathcal{V}_l}\frac{\partial}{\partial t}\int_\Omega \alpha d\mathcal{V} + \frac{1}{\mathcal{V}_l}\oint_{\partial\Omega}\alpha \boldsymbol{u}\cdot\boldsymbol{n}d\mathcal{A} = \frac{1}{\mathcal{V}_l}\int_\Omega \alpha\nabla\cdot\boldsymbol{u}d\mathcal{V}.$$

The right-hand-side is written in non-conservative form and must be solved in a consistent manner with the flux derived in Equation (3.20). As in the work by Johnsen et al. [26], the face value for the volume fraction field is taken to be the area-volume centroid of the cell:

$$\frac{1}{\mathcal{V}_l}\int_\Omega \alpha\nabla\cdot\boldsymbol{u}d\mathcal{V} \approx \frac{\alpha_l}{\mathcal{V}_l}\sum_{f\in\partial\Omega_l} u_{ff}.$$

where α_l denotes the value of alpha at node l. Further, the convective flux term is approximated as:

$$\frac{1}{\mathcal{V}_l}\oint_{\partial\Omega}\alpha \boldsymbol{u}\cdot\boldsymbol{n}d\mathcal{A} \approx \frac{1}{\mathcal{V}_l}\sum_{f\in\partial\Omega_l}\alpha_{f,C}u_{ff},$$

where the computation of $\alpha_{f,C}$ will be detailed next.

CICSAM [35] uses a combination of a controlled down-winding and an up-wind differencing technique to compute $\alpha_{f,C}$. The scheme blends the compressive component, known as Hyper-C and the diffusive component, Ultimate Quickest (UQ) [77] based on the two criteria: boundedness and availability. The boundedness criteria is satisfied through the use of Hyper-C, which is a modification (upper bound value) to the Convection Boundedness Criteria first proposed by Gaskell et al. [78] and re-formulated by Ubbink et al. [35] for explicit flow calculations.

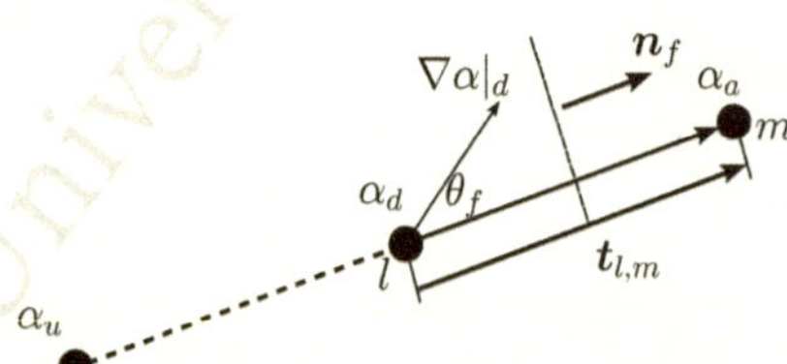

Figure 3.5: Donor-Acceptor method.

In particular, the scheme is based on a donor-acceptor method where the donor node is identified based on the flow direction as illustrated in Figure 3.5:

$$\alpha_d = \begin{cases} \alpha_l, & \text{if } u_{ff} \geq 0, \\ \alpha_m, & \text{otherwise,} \end{cases}$$

where α_d denotes the alpha donor value. The availability criteria is determined using a local cell Courant number, c_d:

$$c_d = \sum_{f \in \partial \Omega_l} \max \left(\frac{u_{ff} \Delta t}{\mathcal{V}_l} \right).$$

To obtain the down-winded Hyper-C component and ensure boundedness, CICSAM first normalises the donor and acceptor values with respect to a projected up-winded value as per the Normalised Variable Diagram [77]. The normalised donor alpha value, $\tilde{\alpha}_d$, is computed via:

$$\tilde{\alpha}_d = \frac{\alpha_d - \alpha_u}{\alpha_a - \alpha_u},$$

where α_a is acceptor alpha value. The availability criteria dictates the amount of fluid to be fluxed at each iteration. This must be less than or equal to the amount of fluid available in the donor cell:

$$\alpha_f c_d \mathcal{V}_d \leq \alpha_d \mathcal{V}_d.$$

Further, controlled down-winding requires that between a donor cell containing both fluids 1 and 2 and the acceptor cell containing only fluid 2, the donor cell must first donate the same fluid contained in the acceptor cell. In other words, only when the donor cell runs out of fluid 2, should it start donating fluid 1. Using these two conditions, the normalised Hyper-C, $\tilde{\alpha}_{cbc}$, is calculated using:

$$\tilde{\alpha}_{cbc} = \begin{cases} \min \left(1, \dfrac{\tilde{\alpha}_d}{c_d} \right), & \text{when } 0 \leq \tilde{\alpha}_d \leq 1, \\ \tilde{\alpha}_d, & \text{when } \tilde{\alpha}_d < 0 \text{ and } \tilde{\alpha}_d > 1. \end{cases}$$

However, this component, on its own, leads to the wrinkling of the interface when the interface is not aligned with the direction of flow. It is therefore blended with the diffusive bounded up-wind differencing scheme, UQ, where:

$$\tilde{\alpha}_{UQ} = \begin{cases} \min\left(\dfrac{8c_d\tilde{\alpha}_d + (1-c_d)(6\tilde{\alpha}_d+3)}{8}, \tilde{\alpha}_{cbc}\right), & \text{when } 0 \le \tilde{\alpha}_d \le 1, \\[2ex] \tilde{\alpha}_d, & \text{when } \tilde{\alpha}_d < 0 \text{ and } \tilde{\alpha}_d > 1, \end{cases}$$

where $\tilde{\alpha}_{UQ}$ denotes the normalised diffusive component of CICSAM. The normalised face value is calculated using:

$$\tilde{\alpha}_f = \zeta_f \tilde{\alpha}_{cbc} + (1-\zeta_f)\tilde{\alpha}_{UQ}$$

where ζ_f is the weighting factor and is expressed as:

$$\zeta_f = \min\left(\frac{\cos(2\theta_f)+1}{2}, 1\right)$$

where θ_f is the angle between gradient in alpha and the tangent vector connecting the donor to the acceptor node:

$$\theta_f = \arccos\left|\frac{(\nabla\alpha^*)_d \cdot \boldsymbol{t}_{l,m}}{|(\nabla\alpha^*)_d|\,||\boldsymbol{t}_{l,m}||}\right|$$

where α^* denotes a smooth filtered field (alpha smooth) and is governed by the following diffusive equation:

$$\frac{\partial\alpha^*}{\partial\tau} - \nabla\cdot\nabla\alpha^* = 0,$$

where τ denotes a sub-time iteration and α^* is computed using:

$$\alpha^{*0} = \alpha^n,$$
$$\alpha_l^{*\tau+1} = \alpha_l^{*\tau} + \frac{\Delta\tau_l}{\mathcal{V}_l}\sum_{f\in\partial\Omega_l}\left(\nabla\alpha^{*\tau}\cdot\boldsymbol{n}\mathcal{A}\right)_f,$$

where $\nabla\alpha^*|_f$ is computed using Central Difference, $\tau \in \mathbb{Z} = \{1,2\}$, $\Delta\tau_l$ is the stable time step for explicit diffusion:

$$\Delta\tau_l = \sigma_{VN}\Delta\mathrm{x}_l^2,$$

and where σ_{VN} denotes the Von-Neumann number set to 0.4. The face value is un-normalised with respect to a blending factor, ξ_f:

$$\alpha_{f,C} = (1-\xi_f)\alpha_d + \xi_f\alpha_a, \tag{3.29}$$

where

$$\xi_f = \frac{\tilde{\alpha}_f - \tilde{\alpha}_d}{1 - \tilde{\alpha}_d},$$

and the nomenclature has been previously defined.

3.6 Summary of Algorithm

Algorithm 1: HLLC with CICSAM

---
1   Initialise α field;
2   Initialise ρ, $\rho\boldsymbol{u}$ and ρE using α;
3   **while** $t < t_{end}$ **do**
4   **for** l *in N nodes* **do**
5   Compute Δt_{min} using Equation (3.8);
6   Reconstruct face primitive variables: ρ_l, $\boldsymbol{u}_l$ and p_l using EOS;
7   **if** $t = 0$ **then**
8   Reconstruct α using Godunov;
9         $\alpha_L = \alpha_l$, $\alpha_R = \alpha_m$;
10   Reconstruct left and right state ($\rho_{k,M}$, $p_{k,M}$, $\boldsymbol{u}_{k,M}$) using Equation (3.5);
11   Compute the face-flux $u_{ff}{}^n$ using Equation (3.20) using the HLLC solver;
12   Using the face-flux u_{ff}^n, compute $\alpha_{f,C}$ using Equation (3.29);
13   **if** α *leads to* $c_k{}^2 < 0$ **then**
14   Reconstruct α_f using Equation (3.5);
15   Compute $\{\frac{1}{\gamma-1}, \frac{\gamma p_\infty}{\gamma-1}\}$ using Equation (3.16) ;
16   Compute $\boldsymbol{F}_f$ using Equation (3.17);
17   Compute curvature using convolution and height functions;
18   Compute surface tension source term using alpha MUSCL;
19   Update the face-flux u_{ff}^{n+1} using Equation (3.20) and conserved variables to ρ^{n+1}, $(\rho\boldsymbol{u})^{n+1}$ and $(\rho E)^{n+1}$ using Equation (3.7) ;

On the first time step, $\alpha_{L/R}$ is reconstructed using Godunov while the primitive variables are reconstructed using MUSCL. Using this initial guess for $\alpha_{L/R}$, HLLC computes a face-flux u_{ff} using Equation (3.20). This face-flux is used in CICSAM to recompute the face value for the volume fraction field. The curvature is then computed. The reconstructed variables are inputs to the HLLC solver where the conservative $\boldsymbol{F}_f$ and face-flux are computed and stored. It is important to note that in the computation of s_*, CICSAM is used for the computation of $\{\frac{1}{\gamma-1}, \frac{\gamma p_\infty}{\gamma-1}\}$ for both the left and right state. Hence the convective flux term is now consistent with the VoF flux. However, for the surface tension term appearing in the contact wave-speed (Equation (3.18)), MUSCL is used for the left and the right state for alpha. This ensures consistency between the pressure gradient operator and the surface tension source term. The benefits of this algorithm is that the use of CICSAM ensures that the volume fraction field is always sharp. This allows curvature to be computed accurately.

Chapter 4

Numerical Test-cases

4.1 Introduction

In this chapter, the developed numerical methodology is assessed and validated. The presented scheme is evaluated using benchmark test-cases found in literature [18, 24, 48]. These test-cases range from 1-D to 2-D, each increasing in complexity from a physics perspective.

First, the compressible component is assessed i.e. the scheme's ability to capture efficiently propagating shocks. This is followed by validating the VoF scheme after which compressible two-phase flow problems are modelled. Finally, surface tension modelling accuracy is evaluated.

For validation purposes, if an analytical solution is available, the spatial order of convergence is computed using the Euclidean (L_1) norm. If such a solution does not exist, the results are compared to existing numerical and experimental data available in literature. Unless otherwise specified, in the interest of VoF accuracy, a CFL= 0.1 is used. In the next section, the 1-D test-cases are presented.

4.2 1-D Test-Cases

The single and two-phase compressible attributes of the proposed solver are evaluated in this section using simple 1-D test-cases. These comprise of the classical Sod-shock tube [79], the interface problem only and the Gas-Liquid Riemann problem [18]. The domain is a simple structured equispaced mesh where outflow boundary conditions are prescribed at each end. Here, the analytical solution is generated by running a very fine mesh ($\Delta x \to 0$) using the original HLLC-MUSCL method and interpolating to find the coarse mesh counter-part (see Appendix A.3).

4.2.1 Sod-Shock tube

A popular test-case [21,32,71,80] for assessing shock capturing accuracy is the Sod-Shock tube [79].

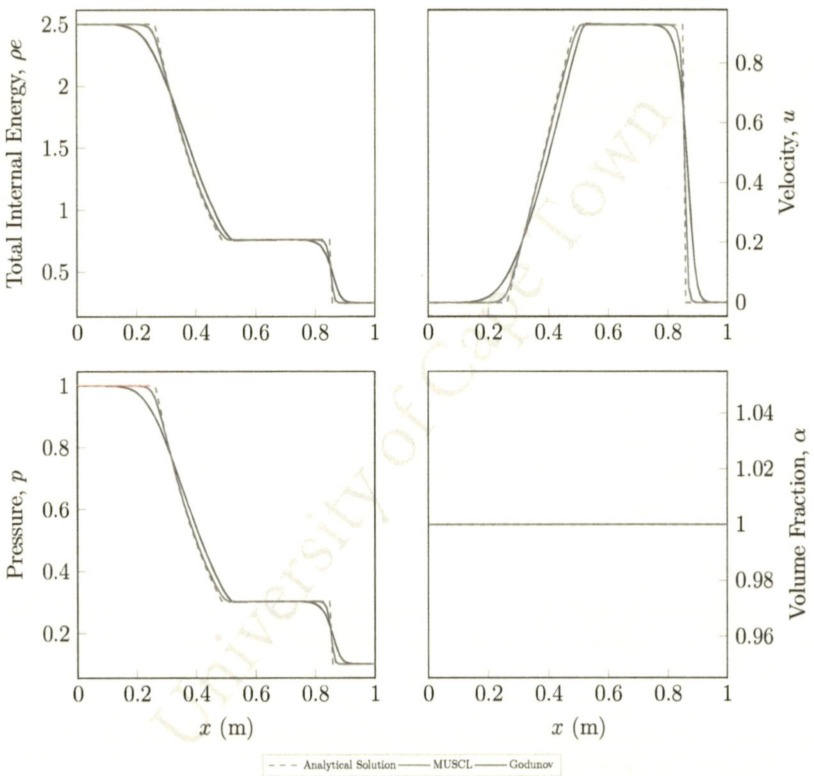

Figure 4.1: Sod Shock Tube - Assessment of Godunov and MUSCL - 100 nodes.

This Riemann problem consists of air at two different states:

$$(\rho, u, p, \gamma, p_\infty) = \begin{cases} (1.0, 0.0, 1.0, 1.4, 0.0), & \text{if } 0 \le x < 0.5, \\ (0.125, 0.0, 0.1, 1.4, 0.0), & \text{if } 0.5 \le x \le 1, \end{cases}$$

where the Stiffened Gas EOS reduces to the ideal gas law. The objective is to demonstrate the ability of the scheme to model single phase compressible flow and illustrate the effect of spatial discretisation on the sharpness of the propagating waves.

To assess this accuracy, MUSCL is compared to the first-order-accurate Godunov method. As depicted in Figure 4.1 by sharp discontinuous regions and linear variations (in the pressure field), the rupture of the diaphragm leads to the propagation of shock and rarefaction waves due to a pressure difference between the left and right state. Here, an excellent agreement is seen between the numerical model and the analytical solution. As expected, HLLC with MUSCL is considerably more accurate than Godunov.

4.2.2   Interface only

Having demonstrated the accuracy of HLLC in modelling sonic characteristics, the VoF component is now assessed. The propagation of an interface in a uniform velocity and pressure field is considered where the solution to this Riemann problem consists of a single contact wave. Two gases with the following material properties:

$$(\rho, u, p, \gamma, p_\infty) = \begin{cases} (1.0, 1.0, 1.0, 1.4, 0.0), & \text{if } 0 \le x < 0.5, \\ (0.125, 1.0, 1.0, 1.2, 0.0), & \text{if } 0.5 \le x \le 1. \end{cases}$$

are passively transported. The objectives are to show the ability of the developed solver to maintain the ECC (Section 3.3.2) as well as CICSAM's superiority over HLLC when applied to the VoF equation. The simulated time is 0.4 s.

As illustrated in Figure 4.2, the pressure and velocity fields are free from spurious oscillations. This indicates that the VoF and energy flux are well-balanced. Moreover, a clear improvement is seen on the advection of the interface when using CICSAM. This is because the latter retains the interface's sharpness in a strictly volume conservative manner by reconstructing a bounded alpha face value using up-wind differencing and controlled down-winding face operators.

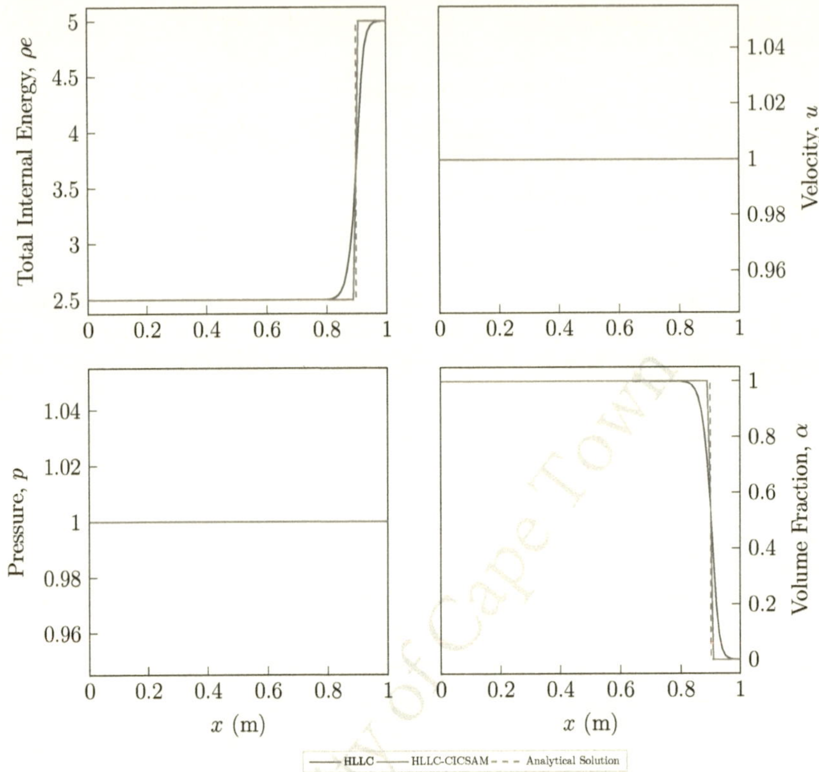

Figure 4.2: Interface only - Assessment of HLLC v. CICSAM - 100 nodes.

4.2.3 Gas-Liquid Shock-tube

The Gas-Liquid Riemann problem is popular [18, 26, 32] to demonstrate the ability of the scheme to capture sonic waves in the presence of a two-phase flow. Here, a high pressure gas interacts with a low pressure liquid of similar density:

$$(\rho, u, p, \gamma, p_\infty) = \begin{cases} (1.241, 0.0, 2.753, 1.4, 0.0)\,, & \text{if } 0 \le x < 0.5, \\ (0.991, 0.0, 3.059 \times 10^{-4}, 5.5, 1.505)\,, & \text{if } 0.5 \le x \le 1. \end{cases}$$

The simulated time is 0.1 s.

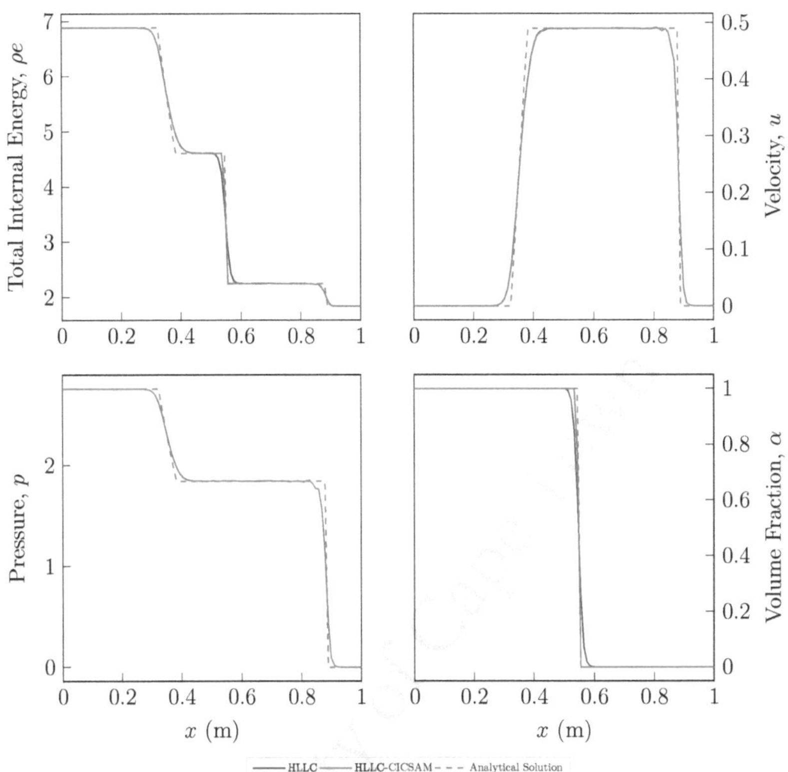

Figure 4.3: Gas-Liquid - Shock-tube results for a 100 node mesh.

Figure 4.3 depicts an overall good agreement between the numerical and analytical solution for both HLLC and CICSAM applied to the VoF equation. CICSAM is again clearly superior at the interface in terms of volume fraction and internal energy. Further, to quantify the formal spatial order of accuracy of the scheme, a mesh independence study is conducted using the L_1 norm of the absolute error. This norm is expressed as:

$$\|\varepsilon\|_1 = \frac{\sum\limits_{l}^{N} |\phi_{nu}(x_l,t) - \phi_{an}(x_l,t)|}{N},$$

where $\phi(x_l,t)$ denotes the value for a variable at position x_l and time t. Finally, ϕ_{nu} and

ϕ_{an} denote the numerical and analytical solution respectively.

From Figure 4.4, for the volume fraction field, it is seen that CICSAM achieves an L_1 norm of at least an order of magnitude lower than that of HLLC. Second-order spatial accuracy is also observed with the proposed CICSAM method when refining from a 400 to 1600 node mesh.

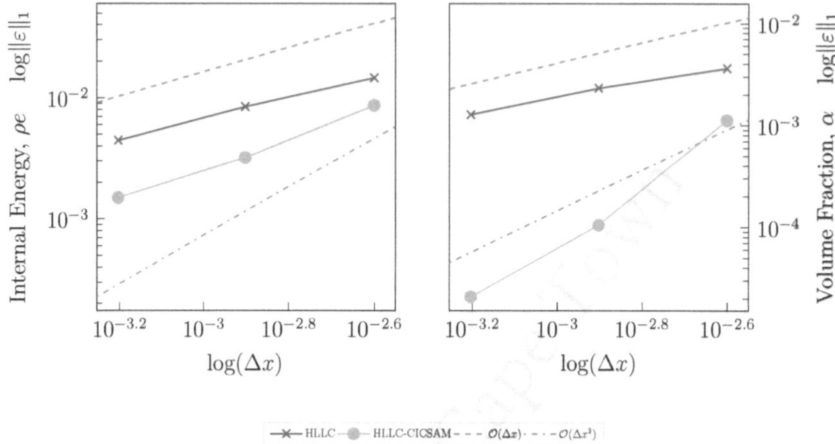

Figure 4.4: Gas-Liquid - Assessment of HLLC v. CICSAM - 400, 800 and 1600 nodes.

4.3   2-D test-cases

In this section, the ability of the scheme to handle multi-dimensional flow is demonstrated. Six test-cases are presented. The first is concerned with evaluating the accuracy of the VoF scheme while the second and third involve two-phase compressible flow problems (without surface tension effects). The final three test-cases are concerned with surface tension modelling accuracy. For all test-cases, VoF initialisation is done using the Arbitrary Grid Initialiser (AGI) by Jones et al. [81]. Unless otherwise stated, slip conditions are prescribed at boundaries.

4.3.1   Advecting bubble in an oblique velocity field

The first test-case is concerned with evaluating the accuracy of the VoF reconstruction method used in combination with HLLC. Two key components are assessed. First, the ability of the scheme to maintain the geometric integrity of the interface. Second, the compatibility of the VoF method with the HLLC solver i.e. its ability to maintain pressure and velocity equilibrium for uniform flow.

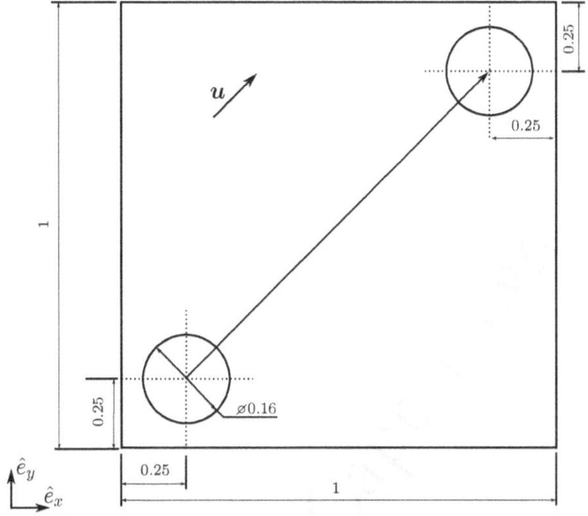

Figure 4.5: Advecting Bubble - Problem Schematic.

As illustrated in Figure 4.5, as per Shyue [48], a circular bubble with radius $R_0 = 0.16$ is centred at the bottom left corner of the domain, $\mathbf{x}_c = (0.25, 0.25)$. The bubble is advected for 0.5 s in the uniform and steady velocity field, $\boldsymbol{u} = 1\hat{\boldsymbol{e}}_x + 1\hat{\boldsymbol{e}}_y$. The initial pressure is set to 1. The material properties are given as:

$$(\rho, \gamma, p_\infty) = \begin{cases} (1.0, 1.4, 0.0), & \text{if } |\mathbf{x} - \mathbf{x}_c| \leq R_0, \\ (0.125, 4, 1.0), & \text{otherwise,} \end{cases}$$

where $\mathbf{x}_c$ denotes the position of the centre of the bubble. Inflow/outflow conditions are enforced at boundaries.

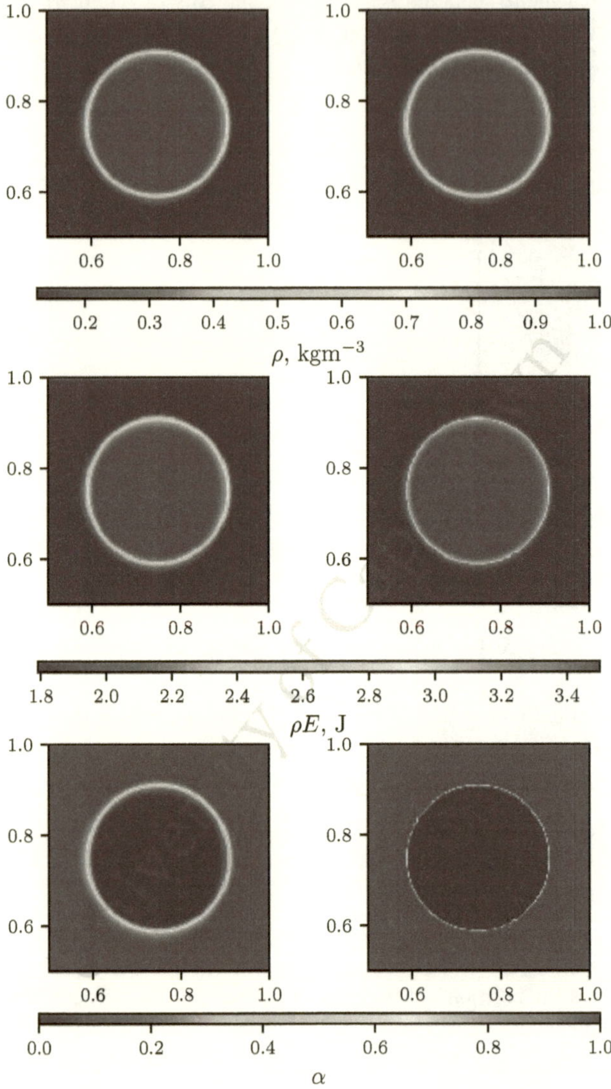

Figure 4.6: Advecting Bubble - HLLC (left) and HLLC-CICSAM (right) on a 512^2 mesh.

Figure 4.6 shows a significant sharper interface with CICSAM than with HLLC. As for the 1-D test-cases, this less diffuse interface leads to an improvement on the qualitative description of the energy field.

To obtain the spatial order of convergence, the calculation is repeated for a set of increasingly finer meshes from 64^2 to 512^2. The L_1 norm of the shape error, ε_g, which measures the absolute error in the reconstruction of the interface, is expressed as:

$$\|\varepsilon_g\|_1 = \frac{\sum\limits_{l}^{N} |\alpha_{nu}(\mathbf{x}_l, t) - \alpha_{an}(\mathbf{x}_l, t)| \, \mathcal{V}_l}{\sum\limits_{l}^{N} \mathcal{V}_l}.$$

Finally, the order of spatial accuracy is computed via:

$$\mathcal{O}(\Delta x) = \frac{\log\|\varepsilon_g\|_1(\frac{\Delta x}{2}) - \log\|\varepsilon_g\|_1(\Delta x)}{\log(\frac{\Delta x}{2}) - \log(\Delta x)}.$$

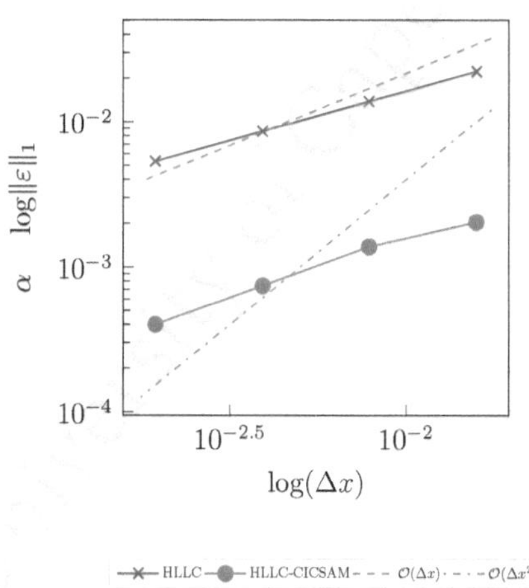

Figure 4.7: Advecting Bubble -L_1 norm error.

As shown in Figure 4.7, the shape error with CICSAM is at least an order of magnitude lower than that of HLLC.

4.3.2 Under-water Explosion

A common 2-D problem to test two-phase compressible flow schemes is that of the under-water explosion [32, 48]. A confined charge is detonated in a rigid tank that is partially filled with water. This charge is represented by a circular region of high pressure, with radius $R_0 = 0.12$ m, which is centred at $\mathbf{x}_c = (0, -0.3)$ m. The domain is rectangular, $(x, y) \in [-2, 2] \times [-1.5, 1]$ m^2.

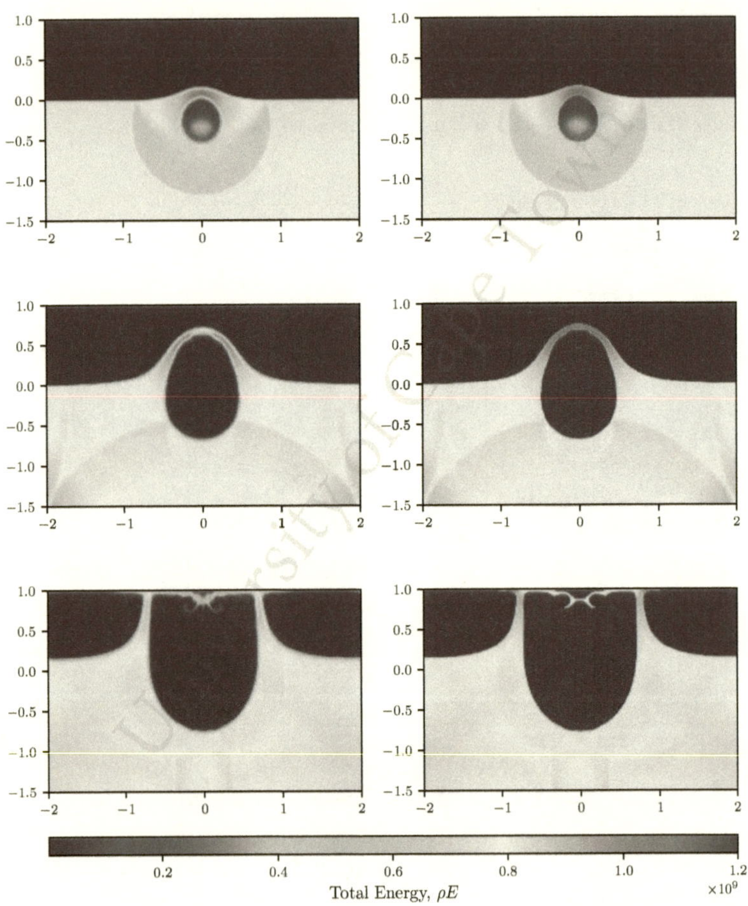

Figure 4.8: Underwater Explosion - HLLC (left) v. HLLC-CICSAM (right).

The free surface separating the air and water layer is at $y = 0$, defining three regions
in the domain:

$$(\rho, p, \gamma, p_\infty) = \begin{cases} \left(1.225\text{kgm}^{-3}, 1.01325 \times 10^5 \text{Pa}, 1.4, 0.0\right), & \text{if } y \geq 0 , \\ \left(1250\text{kgm}^{-3}, 10^9 \text{Pa}, 1.4, 0.0\right), & \text{if } y < 0 \quad \cap \quad |\mathbf{x} - \mathbf{x}_c| \leq R_0, \\ \left(1000\text{kgm}^{-3}, 1.01325 \times 10^5, 4.4, 6.0 \times 10^8 \text{Pa}\right), & \text{if } y < 0 \quad \cap \quad |\mathbf{x} - \mathbf{x}_c| > R_0. \end{cases}$$

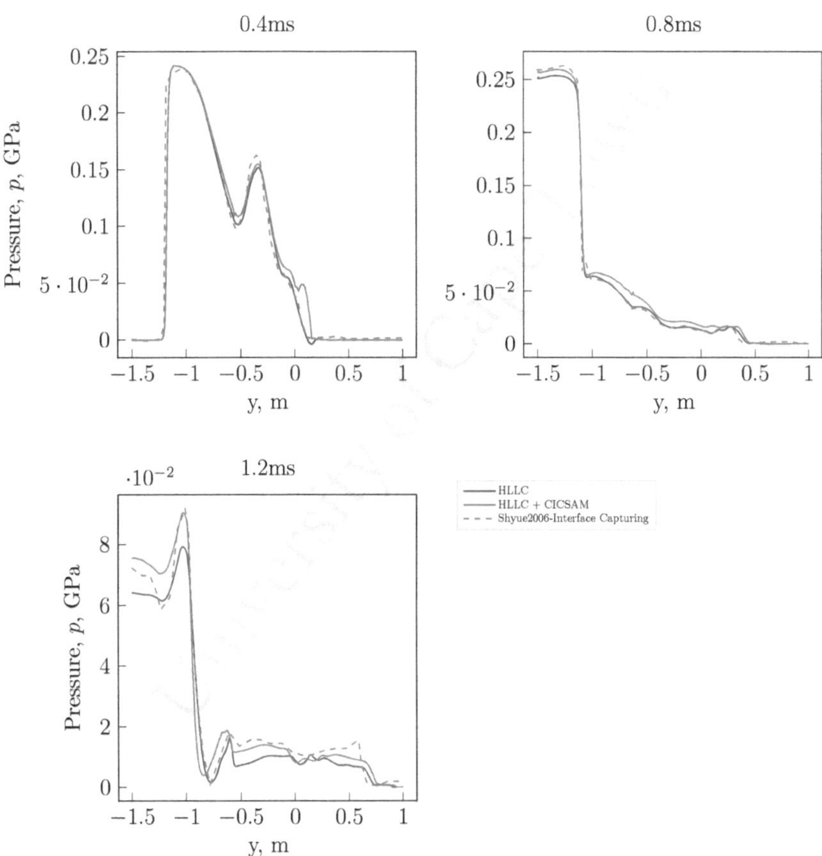

Figure 4.9: Under-water explosion - Pressure.

A structured rectangular mesh with 400×250 nodes is used for this simulation and VOF-HLLC (denoted HLLC) again compared to the proposed scheme (denoted HLLC-CICSAM). Figure 4.8 depicts the evolution of the interface (energy field) at $t = 0.4$ ms (first row), 1.2 ms (second row) and 2.5 ms (third row). At $t = 1.2$ ms, the first shock wave hits the free surface generating a transmitted shock wave in the gas bubble and two rarefaction waves in the liquid. These rarefaction waves interact with those reflected at the lower boundary of the tank. Subsequently, the different transmitted and reflected waves superimpose causing the bubble to change in shape from circular to oval.

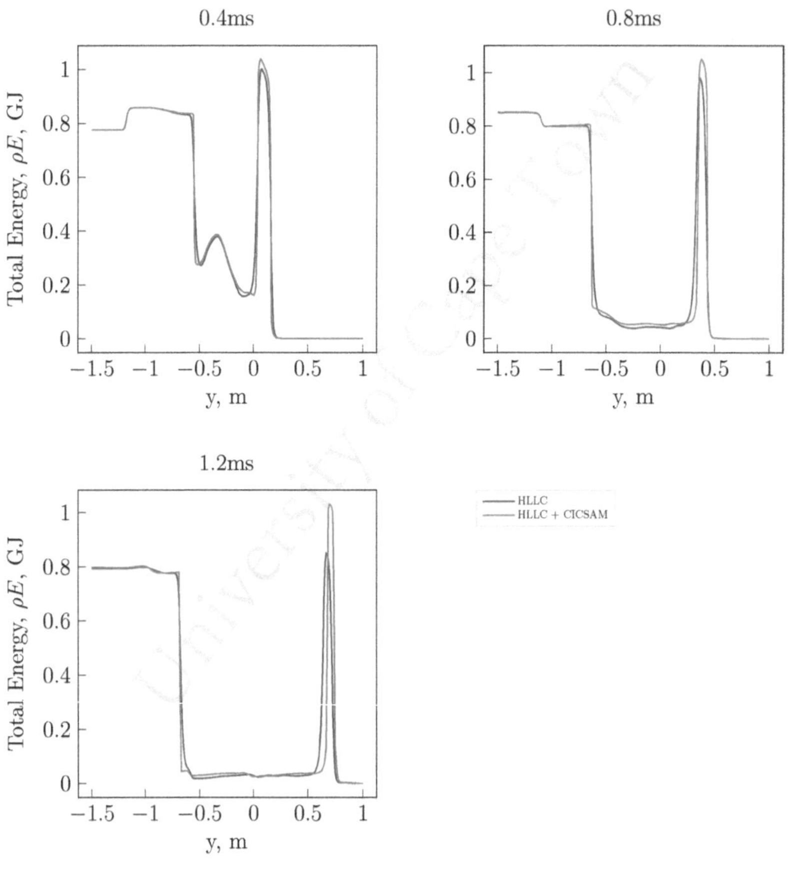

Figure 4.10: Under-water explosion - Total Energy.

The results (Figure 4.8) show flow features that are qualitatively in agreement with those obtained numerically by Shyue [48, Fig. 10]. In particular, the proposed formulation allows for a less diffuse energy field. Further, by plotting the cross-section pressure at $x = 0$ (Figure 4.9), quantitatively, an overall satisfactory agreement is seen with the interface capturing method proposed by Shyue [48]. A higher pressure of up to a factor of 2 is seen in certain regions with HLLC-CICSAM as compared to HLLC. Further, as seen in Figure 4.10, larger energies are recorded with HLLC-CICSAM.

4.3.3 Shock-Bubble Interaction

Another standard test-case for assessing compressible effects is the so-called shock-bubble interaction [26, 32, 48]. A planar Mach 1.22 shock wave, moving from right to left, collides with an R22 gas bubble in air. As shown in Figure 4.11, the gas bubble is of radius $R_0 = 25$ mm and is centred at $\mathbf{x}_c = (225, 44.5)$ mm in a shock-tube of dimensions $(x, y) \in [0, 445] \times [0, 89]$ mm^2.

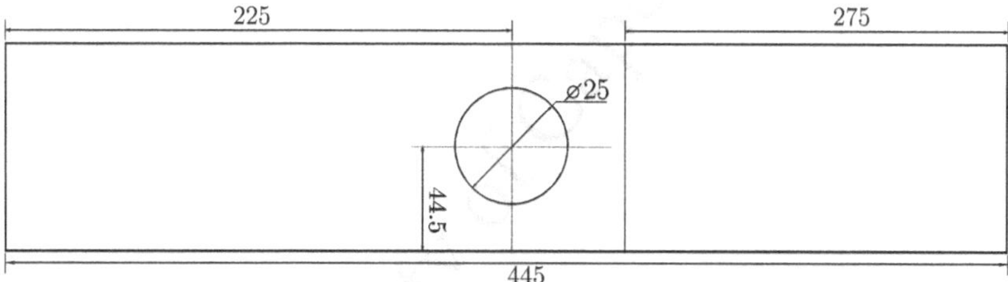

Figure 4.11: Shock-bubble interaction-Initial Set up.

The planar wave is located at $x = 275$ mm, defining pre- and post-shock regions:

$$(\rho, \boldsymbol{u}, p, \gamma, p_\infty) = \begin{cases} \left(1.686\mathrm{kgm}^{-3}, (-113.5, 0)\,\mathrm{ms}^{-1}, 1.59 \times 10^5\mathrm{Pa}, 1.4, 0.0\right), & \text{if } x \leq 275 \text{ mm}, \\ \left(3.863\mathrm{kgm}^{-3}, (0, 0), 1.01325 \times 10^5\mathrm{Pa}, 1.249, 0.0\right), & \text{if } |\mathbf{x} - \mathbf{x}_c| \leq R_0, \\ \left(1.225\mathrm{kgm}^{-3}, (0, 0), 1.01325 \times 10^5\mathrm{Pa}, 1.4, 0.0\right), & \text{if } x > 275\mathrm{mm} \cap \\ & |\mathbf{x} - \mathbf{x}_c| > R_0. \end{cases}$$

The simulation is run on a 3560x356 node mesh at a CFL number of 0.9 for a total time of $1020\mu s$. Figure 4.12 and 4.13 illustrate the evolution of total energy as a function of time. For the same adiabatic coefficient, the speed of sound inside the R22 gas bubble is slower compared to the air around. Hence, the wave generated inside the bubble is a rarefaction wave while the incoming wave is a shock wave. The interaction of these two waves produce two outgoing transmitted waves that get reflected at the boundary and free surfaces. These subsequent waves result in the denser fluid (R22) being decelerated by the lighter fluid (air). These lead to the appearance of Rayleigh-Taylor instabilities as illustrated at $t = 690 - 1020\mu s$.

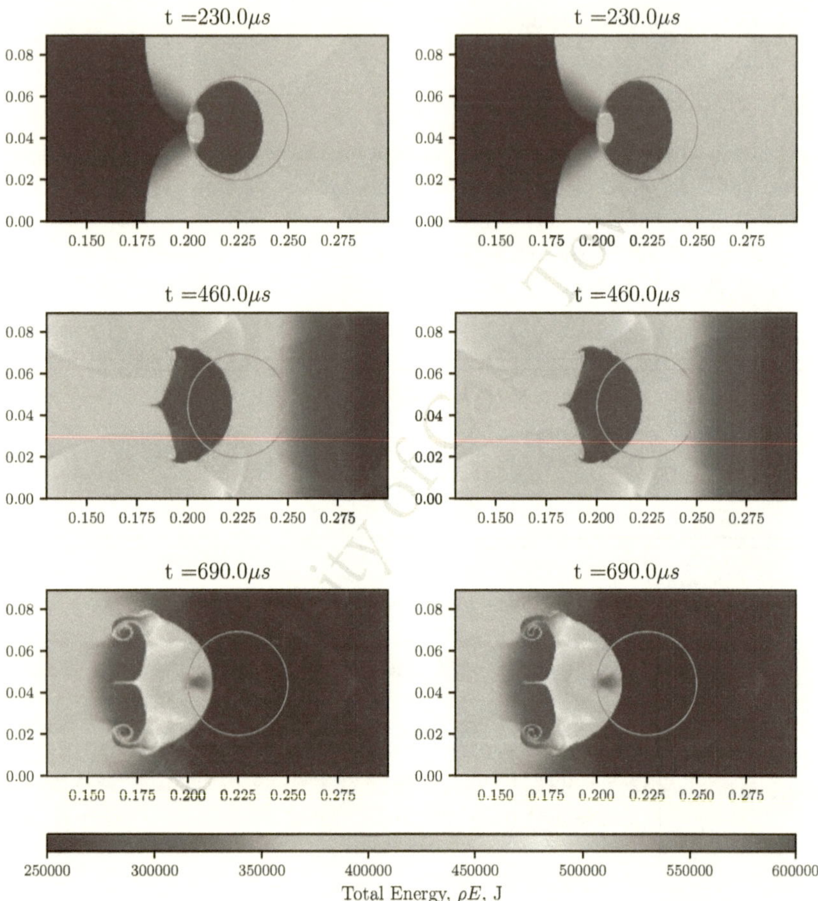

Figure 4.12: Shock-bubble interaction: HLLC (left) v. HLLC-CICSAM (right).

Qualitatively, the plots show flow features that are consistent with numerical results illustrated in [48, Fig. 5] with similar improvements as previously discussed with respect to the VoF and energy field seen when using CICSAM.

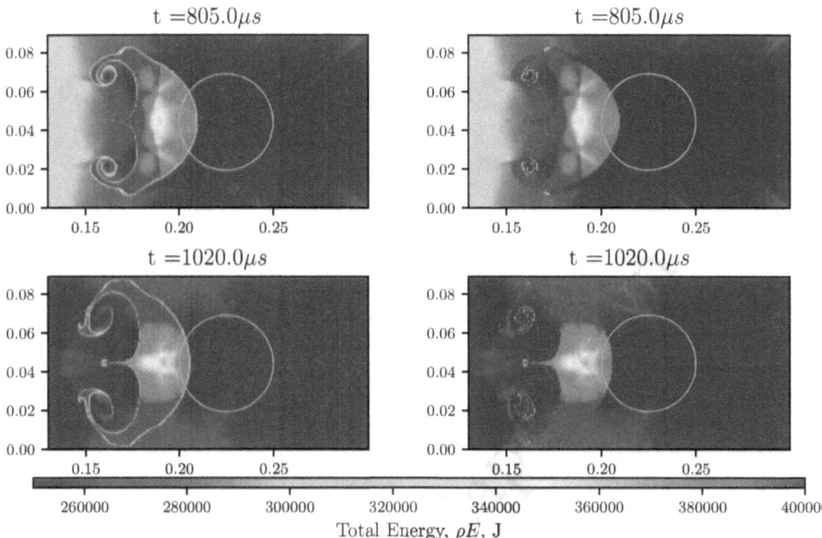

Figure 4.13: Shock Bubble Interaction - HLLC (left) v. HLLC-CICSAM (right).

4.3.4 Spurious currents in a static test-case

Having demonstrated the compressible properties of the scheme, the surface tension component is now assessed. A classical test-case [1, 35, 76, 82] to verify whether the proposed method is well-balanced is that of the static bubble. An ideal gas bubble with a radius of $R_0 = 0.4$ is initialised in a slightly compressible liquid at the centre of a unit domain. The material properties are given as per Fuster et al. [24]:

$$(\rho_1, \gamma_1, p_{\infty_1}) = (1.0, 1.4, 0.0) \text{ and } (\rho_2, \gamma_2, p_{\infty_2}) = (1.0, 7.14, 300),$$

where ρ_1 and ρ_2 denote the phase densities of the gas and liquid respectively. Further, $\sigma = 1$ and there are roughly 12 edges across the radius of the bubble ($\frac{R_0}{\Delta x} = 12.4$). A uniform initial pressure field is set. The simulation is run for 15 s (≈ 179229 steps at $\Delta t \approx 8\mathrm{e}^{-5}$) to allow the momentum to diffuse out for the system to reach steady state.

Here, the infinity norm of the maximum pressure jump is computed using:

$$\|\varepsilon_{\Delta p}\|_\infty = \frac{|\Delta p_{nu} - \Delta p_{an}|}{\Delta p_{an}},$$

where Δp is the difference between the maximum and minimum pressure in the domain, with p_{nu} and p_{an} denoting the numerical and analytical pressure respectively. Further, the velocity is non-dimensionalised with respect to the characteristic inviscid velocity U_σ, which is computed as per Popinet [75]:

$$U_\sigma = \sqrt{\frac{\sigma}{2\rho R_0}}.$$

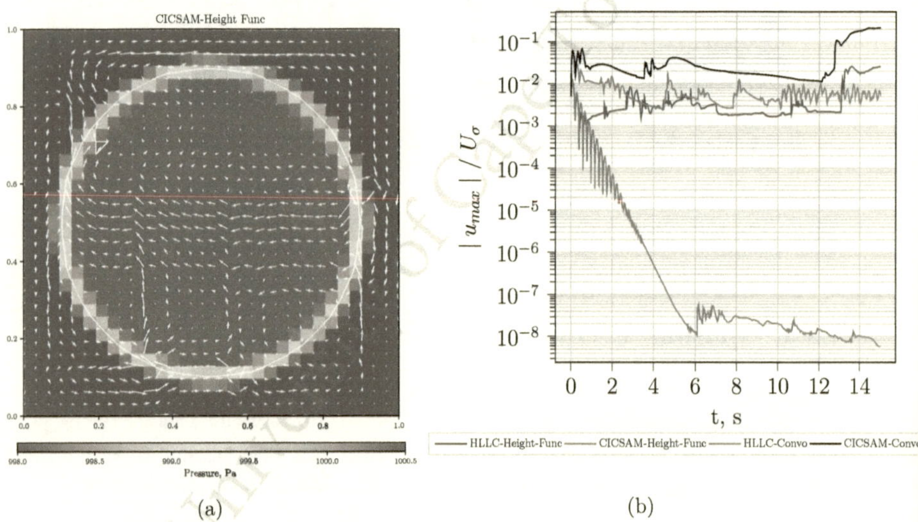

Figure 4.14: Pressure plot (left) and evolution of spurious velocities over time (right).

Table 4.1: Magnitude of max. dimensionless velocity and pressure jump error at $t = 15$ s.

Capturing Scheme	κ	$|\boldsymbol{u}|_{max}/U_\sigma$	$\|\varepsilon(\Delta p)\|_\infty$
HLLC	Height	2.83e-02	4.15e-02
HLLC-CICSAM	Height	6.54e-09	3.20e-03
HLLC	Convolution	5.98e-03	8.74e-03
HLLC-CICSAM	Convolution	2.40e-01	1.27e-01

The results are recorded in Table 4.1. Figure 4.14 (right) illustrates the evolution of the velocity currents over time. The maximum velocity recorded with CICSAM and height functions is of the order of $1e^{-8}$ while that with HLLC is $1e^{-3}$. This difference in magnitude between the two is attributed to MUSCL-more diffused interface with MUSCL. The maximum velocity recorded at steady state for CICSAM with height functions is comparable to the results obtained by Garrick et al. [1] ($1e^{-5}$). This shows that the proposed scheme is well-balanced. The errors when using convolution to compute pressure jump are significantly higher due to a larger error in curvature similar to [83]. In Figure 4.14 (left), the reason for the cells that appear to contain pressure variations while not part of the interface is due to the smeared (algebraic) VoF method employed in this work.

4.3.5 Oscillating bubble

Another widely employed test-case [1, 22, 82, 84, 85] to validate the implementation of surface tension is that of a periodic deformation of a liquid droplet in air in the absence of gravity. The equation for the surface of the droplet is given in polar coordinates by Torres et al. [84]:

$$r\left(\theta\right) = R_0 + \epsilon_r \cos\left(n\theta\right),$$

where n is the integer mode of oscillation. Here for $n = 2$, the above equation can be written in Cartesian coordinates as:

$$\left(x^2 + y^2\right) - \left(\epsilon_r \left[\frac{x^2 - y^2}{x^2 + y^2} + R_0\right]\right)^2 = 0,$$

with $\epsilon_r = 0.01$, $R_0 = 0.8$ and the domain is square of size $[-2, 2]$. The thermodynamic properties for the EOS are given as per Garrick et al. [1]:

$$\left(\gamma_1, p_{\infty_1}\right) = \left(1.4, 0.0\right) \text{ and } \left(\gamma_2, p_{\infty_2}\right) = \left(4.4, 6000\right),$$

where the densities of the gas and liquid are, $\rho_1 = 0.01$ and $\rho_2 = 1$ with $\sigma = 1$. The theoretical small-amplitude inviscid oscillation frequency is then:

$$\omega_n^2 = \frac{(n^3 - n)\,\sigma}{(\rho_1 + \rho_2)\,R_0^3}.$$

The initial velocity is set to zero everywhere.

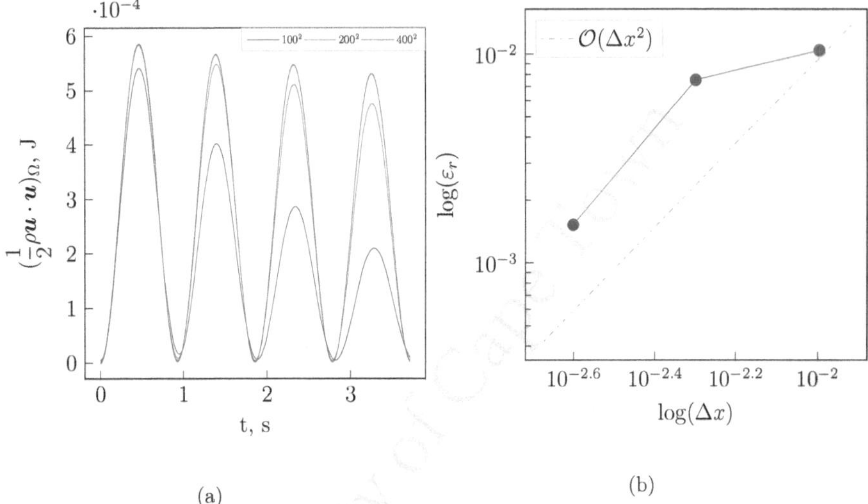

Figure 4.15: Oscillating Bubble - Computed kinetic energy evolution (left) and relative error (right).

The simulation is run for a period of two oscillations corresponding to four complete cycles of kinetic energy at a CFL number of 0.4. Figure 4.15 (left) depicts the evolution of the kinetic energy over time where the damping in the maximum amplitude is due to the numerical dissipation, which is more dominant at low mesh resolutions as seen in [1]. Here, the maximum error in the predicted oscillation frequency on the coarsest mesh is 1.1% and 0.15% on the finest mesh. The results are comparable to those obtained by others [1, 22, 85]. Second-order spatial convergence is recovered when refining from a 200^2 to 400^2 mesh.

4.3.6 Rayleigh-Plesset collapse problem

A popular test-case [24, 86, 87] to evaluate the effect of non-linear terms in multi-phase compressible flow is the so called Rayleigh-Plesset collapse problem [88]. A cylindrical gas bubble (2-D) of radius R_0 at initial pressure $p_b(0)$ collapses in a liquid due to a sudden increase in the surrounding pressure. The liquid is assumed to be incompressible where the evolution of the radius is governed by the 2-D Rayleigh-Plesset model as per [23] (see Appendix A.4 for derivation):

$$\frac{1}{r}\left[\dot{R} + R\ddot{R}\right] - \frac{R^2\dot{R}^2}{r^3} = -\frac{1}{\rho_2}\frac{dp}{dr}. \tag{4.1}$$

Here, R, $\dot{R}$, $\ddot{R}$ are the radius, interface velocity and acceleration of the bubble respectively. By integrating from the bubble radius R on the liquid side at pressure $p_R(t)$ to some finite distance S at a known pressure $p_S(t)$, the second-order non-linear ordinary differential Rayleigh-Plesset equation is recovered:

$$\frac{p_R(t) - p_S(t)}{\rho_2} = \ln\left(\frac{S}{R}\right)\left[\dot{R}^2 + R\ddot{R}\right] + \dot{R}^2\left(\frac{R^2 - S^2}{2S^2}\right). \tag{4.2}$$

As per the Laplace pressure jump condition, the above expression can be re-written as:

$$\frac{p_b(t) - \frac{\sigma}{R} - p_S(t)}{\rho_2} = \ln\left(\frac{S}{R}\right)\left[\dot{R}^2 + R\ddot{R}\right] + \dot{R}^2\left(\frac{R^2 - S^2}{2S^2}\right), \tag{4.3}$$

where the pressure in the bubble, $p_b(t)$, is governed by the poly-tropic gas law:

$$p_b(t) = \left(\frac{R_0}{R}\right)^{2\gamma} p_b(0). \tag{4.4}$$

The unknown pressure $p(r,t)$ in the domain at any time t is:

$$p(r,t) = p_b(t) - \frac{\sigma}{R} - \rho\left[\ln\left(\frac{r}{R}\right)\left[\dot{R}^2 + R\ddot{R}\right] + \dot{R}^2\left(\frac{R^2 - r^2}{2r^2}\right)\right]. \tag{4.5}$$

The initial pressure field is recovered by setting $t = 0$ into the above expression, where $\dot{R}(0) = 0$ and by substituting an expression for the initial acceleration obtained from

Equation (4.2):

$$R_0 \ddot{R}(0) = \frac{p_b(0) - \frac{\sigma}{R_0} - p_S(0)}{\rho_2 \ln\left(\frac{S}{R_0}\right)},$$

(4.6)

into Equation (4.5) as:

$$p(r,0) = p_b(0) - \frac{\sigma}{R_0} - \frac{\ln\left(\frac{r}{R_0}\right)}{\ln\left(\frac{S}{R_0}\right)} \left[p_b(0) - \frac{\sigma}{R_0} - p_S(0)\right].$$

(4.7)

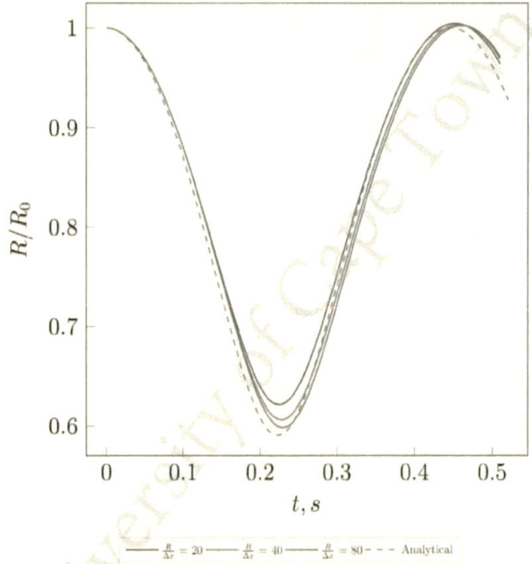

Figure 4.16: Rayleigh-Plesset problem - Square mesh: Predicted bubble radius evolution.

For the purpose of simulating the problem, a bubble of radius $R_0 = 2$ μm is first initialised at the centre of a square domain of size $[-5, 5]$ μm. The driving pressure function at the outer boundary is expressed as:

$$p_S(t) = 200(1 + 0.1\sin(10\omega_c t)),$$

(4.8)

where $\omega_c = 10208967.75$ s^{-1} as per [23, 89]. The material properties are given as per

Fuster et al. [24]:

$$(\rho_1, p_b, \gamma_1, p_{\infty_1}) = (10^{-3}, 100, 1.4, 0.0)$$

$$(\rho_2, p_2, \gamma_2, p_{\infty_2}) = (1.0, 200, 7.14, 3 \times 10^4)$$

with $\sigma = 0.1$.

The Mach number Ma is expressed as

$$Ma = \sqrt{\frac{\Delta p}{\rho_2 c_2^2}}, \tag{4.9}$$

where subscript 2 denotes the liquid phase, $\Delta p = p_2 - p_b$ and for the problem considered $Ma = 2.1 \times 10^{-2}$. At each iteration, a Dirichlet pressure boundary condition (pulse) is set at the outer boundary as per Equation (4.8). The simulation is run at a CFL number of 0.5 for 0.5s.

Figure 4.16 illustrates a reasonable agreement between the numerical and analytical solution which is only valid for finite Mach numbers tending to zero (see Appendix A.5). However, a time lag is noted on the numerical solution with an overshoot on rebound, which is invariant with respect to mesh refinement. This "incorrect" solution is due the square outer boundary whereas the analytical solution involves a circular boundary [87]. To demonstrate this and also to show the applicability of the method to non-orthogonal grids, the calculation is repeated using a curvilinear mesh of radius 5 μm. Since height functions are limited to structured grids, the convolution method is used for the curvature calculation.

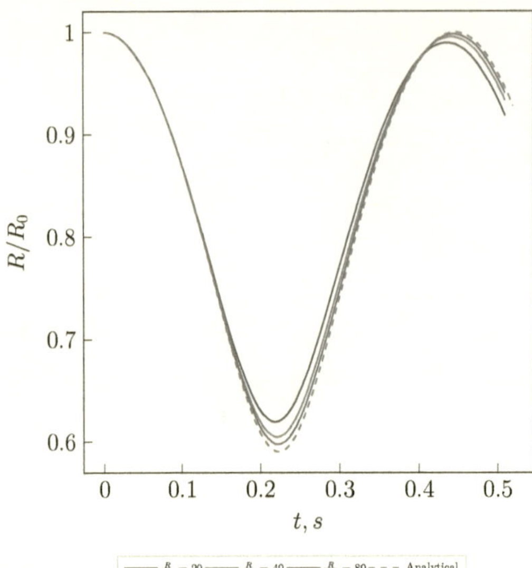

$\frac{R_0}{\Delta x}$	$\varepsilon_r(R_{min})\%$	$\varepsilon_r(T)\%$	$\varepsilon_r(R_r)\%$
20	4.90	2.19	0.001522
40	2.43	0.91	0.001405
80	1.23	0.49	0.001375

Figure 4.17: Rayleigh-plesset problem-Curvilinear mesh (circular domain): Predicted (left) bubble radius evolution and (right) error in minimum radius.

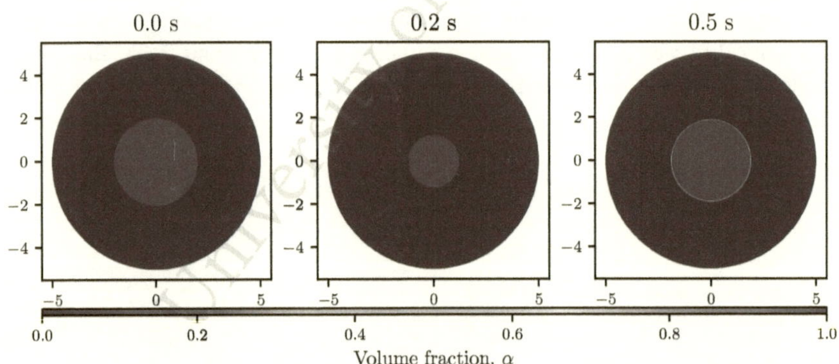

Figure 4.18: Rayleigh-plesset problem-Evolution of volume fraction field.

Figure 4.17 (left) illustrates that the numerical solution now asymptotes to the analytical, correcting the time lag and overshoot. In Table 4.17 (right), the relative error on the minimum radius of collapse, the time period and the amplitude of rebound are recorded. Here, a maximum error of 4.9% is noted on the minimum radius on the coarsest mesh and 1.2% on the finest mesh. Satisfactory convergence with the analytical solution is obtained on the finest mesh ($\frac{R}{\Delta x} = 80$). Figure 4.18 illustrates the evolution of the bubble at the start, mid-way and end time of the simulation. As shown, the CICSAM algorithm allows for sharp capturing of the interface retaining the geometric integrity of the bubble. This test-case demonstrates that the scheme can accurately predict the non-linear collapse of a bubble, indicating that the pressure agrees with the poly-tropic gas law.

Chapter 5

Conclusion

5.1 Summary

In this work, an all Mach HLLC based method capable of accurately modelling two-phase inviscid flow in the presence of surface tension effects was presented. The developed algorithm extended existing numerical work in that a novel approach of coupling the popular HLLC Riemann solver with the algebraic VoF CICSAM method was proposed. The algorithm yields the benefits of both the classical Riemann solver (sharp capture of sonic waves) and that of a typical VoF method (sharp capture of the interface in a strictly volume conservative manner).

The coupling draws on the flexibility of reconstructing primitive rather than conserved variables. In this context, the use of primitive variables enforced the Energy Consistency Criteria, which was set up by the EOS (linking the pressure, energy and VoF flux). Moreover, it was argued that the key criteria to recovering a well-balanced method for surface tension modelling is to enforce consistency between the up-winding technique used for the pressure and alpha field. This is true for the computation of the surface tension term in both the contact wave-speed and the source term.

The proposed methodology was rigorously assessed using a range of test cases from 1-D to 2-D. First, simple 1-D test cases for single and two-phase flow were presented to assess sharp shock capturing ability. The algorithm showed clear improvements of at least an order of magnitude on the capture of the interface when using CICSAM as opposed to HLLC. The less diffuse interface lead to an improvement in the representation of the energy field. To show the compatibility of the HLLC solver with the VoF method in 2-D, the advection of a bubble in an oblique velocity field was presented. Clear improvements were seen on the geometric accuracy of the interface.

To test the ability of the scheme to handle violent (high Mach numbers) flows, the method was assessed using the popular under-water explosion. Similar improvements on the volume fraction and energy field were seen when using CICSAM v. HLLC for VoF. Overall, the pressure field was observed to be in satisfactory agreement to that obtained by Shyue [48]. Finally, to provide further qualitative comparison between HLLC and HLLC-CICSAM, the shock-bubble interaction test case was presented. The results showed flow features consistent with those obtained in literature [48].

To verify that the scheme works for surface tension, two test cases were presented. First, the classic static bubble and then the periodic deformation of a circular droplet. For the former, the velocity spurious currents were damped to the order of $1e^{-8}$ with HLLC-CICSAM and height functions. For the second test case, second-order spatial convergence was obtained on the computation of the periodic deformation. To show the robustness of the scheme and its applicability to non-orthogonal meshes, the popular Rayleigh-Plesset collapse problem was considered. Again, an excellent agreement with the analytical solution for the time period, minimum radius of collapse and amplitude of rebound was seen.

5.2 Recommendations and future work

The proposed algorithm is suitable for modelling inviscid compressible flow on arbitrary meshes, but the following recommendation is made with regard to future work:

- Extension to a 3-D formulation: Since the expressions for the HLLC and VoF (CICSAM) flux presented in this work are written for the general edge and the discretisation has been done in an edge-wise manner, the described mathematics is fully applicable to a 3-D formulation. In fact, the author's code is fully 3-D capable. The reason for excluding any 3-D results was due to the longer computational times and time constraints on completing the Masters' degree.

- Inclusion of viscous effects: The current formulation will need to be extended and further work will have to be done to ensure temporal stability when including viscous effects. In this regard, for different temperatures, the viscosity may be re-computed via the Sutherland law [90, p. 18].

- Improving the spatial accuracy: On certain cases, low spatial convergence rates are observed. This is mainly due to the fact that MUSCL is dissipative. A higher order method such as the fifth order Weighted-Essentially-Non-Oscillatory (WENO) scheme could be used for the primitive variable reconstruction.

- Solver: A comparison between the semi-implicit method proposed by Fuster et al. [24] and the HLLC-CICSAM method proposed in this work could be conducted. In particular, differences in the performance of the solver between initializing the discontinuity aligned with the grid or placing the discontinuity between two cell faces could be evaluated.

- VoF method: A geometric VoF method could be implemented in a similar manner to that presented in this work and a comparison on the accuracy could be drawn with respect to algebraic VoF method.

- Curvature: A second-order method for computing curvature on unstructured meshes could also be implemented to obtain a fully second-order scheme.

- Hyperbolicity: When running certain simulation test-cases such as the under-water explosion for longer periods of time, imaginary eigenvalues (non-physical acoustic velocities) resulted. This may be due a loss of hyperbolicity as per [61] and is to be further investigated in future work.

Appendices

Appendix A

A.1   Derivation of thermal EOS

The expression for the thermal EOS is derived as follows. The internal energy is first written as a function of pressure and specific volume v:

$$e\left(v,T\right) = \left[\frac{p\left(v,T\right)}{\gamma - 1} + \frac{\gamma p_{\infty}}{\gamma - 1}\right] v. \tag{A.1}$$

By differentiating the internal energy with respect to (w.r.t.) the temperature at constant volume, the following relationship is obtained:

$$\left(\frac{\partial e}{\partial T}\right)_{v} = \frac{v}{\gamma - 1}\left(\frac{\partial p}{\partial T}\right)_{v}. \tag{A.2}$$

The internal energy is also differentiated w.r.t. specific volume at constant temperature, yielding:

$$\left(\frac{\partial e}{\partial v}\right)_{T} = \frac{v}{\gamma - 1}\left(\frac{\partial p}{\partial v}\right)_{T} + \frac{p\left(v,T\right) + \gamma p_{\infty}}{\gamma - 1}. \tag{A.3}$$

Maxwell's rule relates the internal energy to the temperature, pressure and specific volume as follows:

$$\left(\frac{\partial e}{\partial v}\right)_{T} = T\left(\frac{\partial p}{\partial T}\right)_{v} - p. \tag{A.4}$$

By substituting Equation (A.3) into Equation (A.4), the following expression is obtained:

$$\left(\frac{\partial p}{\partial v}\right)_{T} = \frac{\left(\gamma - 1\right)T}{v}\left(\frac{\partial p}{\partial T}\right)_{v} - \frac{\gamma\left(p\left(v,t\right) + p_{\infty}\right)}{v}. \tag{A.5}$$

Further, substituting for the specific heat capacity at constant volume, $c_v = \left(\frac{\partial e}{\partial T}\right)_v$ into Equation (A.2), leads to:

$$\left(\frac{\partial p}{\partial T}\right)_v = \frac{(\gamma - 1)\,c_v}{v}. \tag{A.6}$$

Integrating with respect to T yields:

$$p\,(v, T) = \frac{(\gamma - 1)\,c_v T}{v} + \beta\,(v). \tag{A.7}$$

Next differentiating the above with respect to the volume at constant temperature, leads to:

$$\left(\frac{\partial p}{\partial v}\right)_T = -\frac{(\gamma - 1)\,c_v T}{v^2} + \frac{d\beta}{dv}, \tag{A.8}$$

Substituting Equations (A.6) and (A.7) into (A.5) leads to the following expression:

$$\left(\frac{\partial p}{\partial v}\right)_T = -\frac{(\gamma - 1)\,c_v T}{v^2} - \frac{\gamma}{v}\,(\beta\,(v) + p_\infty). \tag{A.9}$$

Substituting Equation (A.9) into Equation (A.8) yields:

$$\frac{d\beta}{dv} + \frac{\gamma}{v}\beta\,(v) = -\frac{\gamma}{v}p_\infty \tag{A.10}$$

Using an integrating factor such as $e^{\int \frac{\gamma}{v}dv}$, an expression for $\beta\,(v)$ is obtained as follows:

$$\beta\,(v) = \frac{A}{v^\gamma} - p_\infty, \tag{A.11}$$

where A denotes the constant of integration.

Then, substituting Equation (A.11) into Equation (A.7), an expression for $p\,(v, T)$ is derived as:

$$p\,(v, T) = \frac{(\gamma - 1)\,c_v T}{v} + \frac{A}{v^\gamma} - p_\infty. \tag{A.12}$$

Using a reference state (p_0, T_0, v_0), an expression for A is obtained as:

$$A = \left[p_0 + p_\infty - \frac{(\gamma - 1)\,c_v T_0}{v_0}\right] v_0^\gamma. \tag{A.13}$$

Re-expressing the above equation in terms of the reference density ρ_0 yields:

$$A = \left[p_0 + p_\infty - (\gamma - 1)\, c_v T_0 \rho_0\right] \rho_0^{-\gamma}. \tag{A.14}$$

Finally, substituting Equation (A.14) into Equation (A.7) and re-expressing in terms of density, the equation for temperature reads:

$$T = \frac{1}{\rho c_v} \left[\frac{p + p_\infty - A\rho^\gamma}{\gamma - 1}\right]. \tag{A.15}$$

where the nomenclature has been previously defined.

A.2 Derivation of HLLC intermediate star-state

Consider the general formulation for the HLLC solver:

$$\boldsymbol{F}_{*_k} = \boldsymbol{F}_k + s_k \left(\boldsymbol{U}_{*_k} - \boldsymbol{U}_k\right). \tag{A.16}$$

By substituting for known quantities $\boldsymbol{U}_k$ and $\boldsymbol{F}_k$, an expression for the star-state region is obtained. First, for the continuity equation, this reads:

$$\rho_{*_k} \boldsymbol{u}_{*_k} \cdot \boldsymbol{n} = \rho_k \boldsymbol{u}_k \cdot \boldsymbol{n} + s_k \left(\rho_{*_k} - \rho_k\right),$$
$$\rho_{*_k} s_* = \rho_k \boldsymbol{u}_k \cdot \boldsymbol{n} + s_k \left(\rho_{*_k} - \rho_k\right).$$

Hence

$$\rho_{*_k} \left(s_k - s_*\right) = \rho_k \left(s_k - \boldsymbol{u}_k \cdot \boldsymbol{n}\right), \tag{A.17}$$
$$\rho_{*_k} = \frac{s_k - \boldsymbol{u}_k \cdot \boldsymbol{n}}{s_k - s_*} \rho_k, \tag{A.18}$$

and for the momentum equation, this is expanded as follows:

$$\rho_{*_k} \boldsymbol{u}_{*_k} \boldsymbol{u}_{*_k} \cdot \boldsymbol{n} + p_{*_k} \boldsymbol{n} = \rho_k \boldsymbol{u}_k \boldsymbol{u}_k \cdot \boldsymbol{n} + p_k \boldsymbol{n} + s_k \left(\rho_{*_k} \boldsymbol{u}_{*_k} - \rho_k \boldsymbol{u}_k\right),$$
$$p_{*_k} \boldsymbol{n} = \rho_k \boldsymbol{u}_k \boldsymbol{u}_k \cdot \boldsymbol{n} + p_k \boldsymbol{n} + \left(s_k - s_*\right) \rho_{*_k} \boldsymbol{u}_{*_k} - s_k \rho_k \boldsymbol{u}_k, \tag{A.19}$$

From consistency Equation (3.9), the following expression must hold:

$$\boldsymbol{u}_{*_k} = \boldsymbol{u}_{*n,k} + \boldsymbol{u}_t,$$

where $\boldsymbol{u}_{*n,k}$ and $\boldsymbol{u}_t$ are the normal and tangential velocities respectively. Hence

$$\boldsymbol{u}_{*_k} = \boldsymbol{u}_{*n,k} + \boldsymbol{u}_t,$$
$$\boldsymbol{u}_{*_k} = s_* \boldsymbol{n} + \boldsymbol{u}_k - (\boldsymbol{u}_k \cdot \boldsymbol{n}) \, \boldsymbol{n},$$
$$\boldsymbol{u}_{*_k} = \boldsymbol{u}_k + (s_* - \boldsymbol{u}_k \cdot \boldsymbol{n}) \, \boldsymbol{n}. \tag{A.20}$$

Substituting Equations (A.18) and (A.20) into Equation (A.19), an expression for the intermediate pressure, p_{*_k}, is derived as follows:

$$p_{*_k} = p_k + \rho_k \left(s_k - \boldsymbol{u}_k \cdot \boldsymbol{n} \right) \left(s_* - \boldsymbol{u}_k \cdot \boldsymbol{n} \right) \tag{A.21}$$

Substituting Equation (A.21) into Equation (A.19), an expression for the intermediate momentum is obtained:

$$\rho_{*_k} \boldsymbol{u}_{*_k} = \frac{s_k - \boldsymbol{u}_k \cdot \boldsymbol{n}}{s_k - s_*} \left[\rho_k \left[\boldsymbol{u}_k + (s_* - \boldsymbol{u}_k \cdot \boldsymbol{n}) \, \boldsymbol{n} \right] \right] \tag{A.22}$$

Similarly, for the energy equation, the HLLC flux is written as:

$$\rho_{*_k} E_{*_k} \boldsymbol{u}_{*_k} \cdot \boldsymbol{n} + p_{*_k} \boldsymbol{u}_{*_k} \cdot \boldsymbol{n} = \rho_k E_k \boldsymbol{u}_k \cdot \boldsymbol{n} + s_k \left(\rho_{*_k} E_{*_k} - \rho_k E_k \right) \tag{A.23}$$

Substituting the consistency Equation (3.9) and Equation (A.21) into Equation (A.23) yields the expression for the intermediate star-state energy term as:

$$\rho_{*_k} E_{*_k} = \frac{s_k - \boldsymbol{u}_k \cdot \boldsymbol{n}}{s_k - s_*} \rho_k \left[E_k + (s_* - \boldsymbol{u}_k \cdot \boldsymbol{n}) \left[s_* + \frac{p_k}{\rho_k \left(s_k - \boldsymbol{u}_k \cdot \boldsymbol{n} \right)} \right] \right]. \tag{A.24}$$

Finally, using consistency Equation (3.10), the following expression for the contact wave speed is obtained:

$$s_* = \frac{p_L - p_R + \rho_R \boldsymbol{u}_R \cdot \boldsymbol{n} \left(s_R - \boldsymbol{u}_R \cdot \boldsymbol{n} \right) - \rho_L \boldsymbol{u}_L \cdot \boldsymbol{n} \left(s_L - \boldsymbol{u}_L \cdot \boldsymbol{n} \right) - \sigma \kappa \left(\alpha_L - \alpha_R \right)}{\rho_R \left(s_R - \boldsymbol{u}_R \cdot \boldsymbol{n} \right) - \rho_L \left(s_L - \boldsymbol{u}_L \cdot \boldsymbol{n} \right)} \tag{A.25}$$

where s_k is computed as per Einfeldt et al. [74].

A.3 Analytical solution for 1-D test cases

Consider a point l on a coarse 1-D Cartesian mesh with coordinate x_l for a property ϕ. Say that this point lies between two coordinates points x^F_{p-1} and x^F_p on a very fine Cartesian mesh. For an arbitrary variable ϕ, the analytical solution is interpolated using:

$$\phi^a_l = \phi^F_{p-1} + SF(\phi^F_p - \phi^F_{p-1}),$$
(A.26)

where superscript F denotes the value of ϕ on the fine mesh with:

$$SF = \frac{x_l - x^F_{p-1}}{x^F_p - x^F_{p-1}}.$$
(A.27)

where ϕ^F is computed using HLLC-MUSCL on a very fine mesh for which $\Delta x \approx 1e^{-5}$.

A.4 Derivation of 2-D Rayleigh-Plesset Equation

The derivation for the Rayleigh-Plesset equation detailed in this work draws from the following literature [23, 52, 88]. First, the liquid is assumed to be incompressible. Hence, the divergence free condition requires that:

$$\nabla \cdot \boldsymbol{u} = 0.$$

Then, the flow is assumed to be spherically symmetric and thus the velocity is only a function of the radial coordinate, r, and time t:

$$\frac{du}{dr} = 0 \implies u = u(r,t),$$

where u denotes the radial component of the velocity. Further, the velocity field is parametrised and assumed to follow a linear relationship with respect to the distance r from the centre of the bubble:

$$u(r,t) = \frac{g(t)}{r},$$
(A.28)

where $g(t)$ denotes a time-dependent function. Therefore, the velocity at the bubble interface (two-phase) is such that $u(R,t) = \frac{dR}{dt} = \dot{R}$ and hence:

$$u(R,t) = \frac{g(t)}{R}, \qquad \text{and} \qquad R\dot{R} = g(t). \tag{A.29}$$

Substituting Equation (A.29) into Equation (A.28) yields:

$$u = \frac{R\dot{R}}{r}. \tag{A.30}$$

Using the momentum equation for an inviscid incompressible homogeneous two-phase flow, an expression relating the pressure and radial velocity is obtained as follows:

$$\frac{du}{dt} + u\frac{du}{dr} + \frac{1}{\rho_2}\frac{dp}{dr} = 0. \tag{A.31}$$

Finally, substituting Equation (A.30) into (A.31):

$$\frac{d}{dt}\left(\frac{R\dot{R}}{r}\right) + \frac{R\dot{R}}{r}\frac{d}{dr}\left(\frac{R\dot{R}}{r}\right) = -\frac{1}{\rho_2}\frac{dp}{dr}, \tag{A.32}$$

yields the second-order non-linear differential equation:

$$\frac{1}{r}\left[\dot{R} + R\ddot{R}\right] - \frac{R^2\dot{R}^2}{r^3} = -\frac{1}{\rho_2}\frac{dp}{dr}. \tag{A.33}$$

where the nomenclature has been previously defined.

A.5 Runge-Kutta fourth-order - Rayleigh-Plesset model

Equation (4.2) may be expressed as:

$$\frac{\left(\frac{R_0}{R}\right)^{2\gamma} p_b(0) - p_S(t)}{\rho} = \ln\left(\frac{S}{R}\right)\left[\dot{R}^2 + R\ddot{R}\right] + \dot{R}^2\left(\frac{R^2 - S^2}{2S^2}\right).$$

The above equation is a second-order non-linear Ordinary Differential Equation (ODE) and is solved as two first-order non-linear ODE w.r.t. v and its derivative $\dot{v}$ as:

$$R = R,$$

$$v = \dot{R} \implies \dot{R} = v \implies f_1\left(t, R, v\right),$$

$$\dot{v} = \ddot{R} = \frac{1}{R}\left[\frac{\dfrac{p_b(0)\left(\frac{R_0}{R}\right)^{2\gamma} - \frac{\sigma}{R} - p_S(t)}{\rho_2} - \dot{R}^2\left(\frac{R^2 - S^2}{2S^2}\right)}{\ln\left(\frac{R}{S}\right)} - \dot{R}^2\right] \implies f_2\left(t, R, v\right).$$

Hence, R and v are updated via a fourth-order Runge-kutta method:

$$R^{n+1} = R^n + \frac{\Delta t}{6}\left(\mathcal{F}_{1_r} + 2\mathcal{F}_{2_r} + 2\mathcal{F}_{3_r} + \mathcal{F}_{4_r}\right),$$

$$v^{n+1} = v^n + \frac{\Delta t}{6}\left(\mathcal{F}_{1_v} + 2\mathcal{F}_{2_v} + 2\mathcal{F}_{3_v} + \mathcal{F}_{4_v}\right),$$

with

$$\mathcal{F}_{1_r} = f_1\left(t, R^n, v^n\right),$$

$$\mathcal{F}_{2_r} = f_1\left(t + \frac{\Delta t}{2}, R^n + \frac{\Delta t}{2}\mathcal{F}_{1_r}, v^n + \frac{\Delta t}{2}\mathcal{F}_{1_v}\right),$$

$$\mathcal{F}_{3_r} = f_1\left(t + \frac{\Delta t}{2}, R^n + \frac{\Delta t}{2}\mathcal{F}_{2_r}, v^n + \frac{\Delta t}{2}\mathcal{F}_{2_v}\right),$$

$$\mathcal{F}_{4_r} = f_1\left(t + \Delta t, R^n + \Delta t\mathcal{F}_{3_r}, v^n + \Delta t\mathcal{F}_{3_v}\right),$$

and where $\mathcal{F}_{n_v}$ is computed as per the above except with f_1 replaced by f_2.

A.6 Ethics Clearance

ETHICS APPLICATION FORM

Please Note:

Any person planning to undertake research in the Faculty of Engineering and the Built Environment (EBE) at the University of Cape Town is required to complete this form **before** collecting or analysing data. The objective of submitting this application *prior* to embarking on research is to ensure that the highest ethical standards in research, conducted under the auspices of the EBE Faculty, are met. Please ensure that you have read, and understood the **EBE Ethics in Research Handbook** (available from the UCT EBE, Research Ethics website) prior to completing this application form: http://www.ebe.uct.ac.za/ebe/research/ethics1

APPLICANT'S DETAILS		
Name of principal researcher, student or external applicant		Muhammad Yusufali OOMAR
Department		Mechanical Engineering
Preferred email address of applicant:		omrmuh001@myuct.ac.za
If Student	Your Degree: e.g., MSc, PhD, etc.	MSc (Eng) in mechanical Engineering
	Credit Value of Research: e.g., 60/120/180/360 etc.	180
	Name of Supervisor (if supervised):	Arnaud Malan
If this is a researchcontract, indicate the source of funding/sponsorship		
Project Title		A Volume Of fluid (VOF) based HLLC method for multiphase compressible flow method

I hereby undertake to carry out my research in such a way that:

- there is no apparent legal objection to the nature or the method of research; and
- the research will not compromise staff or students or the other responsibilities of the University;
- the stated objective will be achieved, and the findings will have a high degree of validity;
- limitations and alternative interpretations will be considered;
- the findings could be subject to peer review and publicly available; and
- I will comply with the conventions of copyright and avoid any practice that would constitute plagiarism.

APPLICATION BY	Full name	Signature	Date
Principal Researcher/ Student/External applicant	Muhammad Yusufali OOMAR		6/02/2020
SUPPORTED BY	Full name	Signature	Date
Supervisor (where applicable)	Arnaud Malan		6/2/2020

APPROVED BY	Full name	Signature	Date
HOD (or delegated nominee) Final authority for all applicants who have answered NO to all questions in Section 1; and for all Undergraduate research (including Honours).	T Bello-Ochende		13/08/2020
Chair: Faculty EIR Committee For applicants other than undergraduate students who have answered YES to any of the questions in Section 1.			

Page 1 of

www.ingramcontent.com/pod-product-compliance
Lightning Source LLC
LaVergne TN
LVHW041745190726
843493LV00008B/2455